FEDERAL EMERGENCY MANAGEMENT AGENCY
UNITED STATES FIRE ADMINISTRATION
NATIONAL FIRE DATA CENTER

Firefighter Fatality Summaries
1990 - 2000

April 2002

Prepared by:

TriData Corporation
1000 Wilson Boulevard
Arlington, VA 22209

U.S. Fire Administration
Mission Statement

As an entity of the Federal Emergency Management Agency, the mission of the U.S. Fire Administration is to reduce life and economic losses due to fire and related emergencies through leadership, advocacy, coordination, and support. We serve the National independently, in coordination with other Federal agencies and in partnership with fire protection and emergency service communities. With a commitment to excellence, we provide public education, training, technology, and data initiatives.

Introduction

Fire departments in the United States respond to an average of 2 million fires each year. The men and women who respond to these fires are at extreme risk when performing their duties. Over an 11-year period from 1990 to 2000, more than 1 million firefighters have been injured, and 1,068 have lost their lives.

This report focuses on the fatalities suffered by firefighters and other officials engaged in the suppression of fires over this 11-year span. The bulk of this report is in chronological order by the date of the incident that caused the death, and it reflects the individual's name, affiliation, and department. Summaries of the incidents themselves were unavailable for the years 1990–1993, but they are included for the next 7 years. An alphabetized list of fatalities begins on page 105.

All of the data presented were compiled from the U.S. Fire Administration's firefighter fatality reports. Some additional data were collected by TriData Corporation during the report's preparation.

This report was produced by TriData Corporation, Arlington, Virginia, for the National Fire Data Center, U.S. Fire Administration, under Contract Number EME–2000–DO–0396.

* * *

The abbreviations and acronyms used in this report are defined below:

AED	automatic external defibrillator
ALS	advanced life support
ARFF	airport rescue fire fighting (vehicle)
BLEVE	boiling liquid expanding vapor explosion
BLS	basic life support
CPR	cardiopulmonary resuscitation
CVA	cardiovascular accident
EMS	emergency medical service
FAO	fire apparatus operator
MVA	motor vehicle accident
NIOSH	National Institute for Occupational Safety and Health
PASS	Personal Alarm Safety System
SCBA	self-contained breathing apparatus
SUV	sports utility vehicle
USDA	U.S. Department of Agriculture

Fatality Summary: 1990–1993

Date of Incident	Full Name	Affiliation	State	Department
1/1/90	Joseph M. Wilt	Volunteer	IA	West Burlington Fire Dept
1/1/90	William T. Klein	Volunteer	IA	West Burlington Fire Dept
1/10/90	James E. White	Career	TX	Carswell AFB Fire Dept
1/15/90	Carrol D. Marvel	Volunteer	AL	Valley Vol Fire Dept
1/18/90	Francis N. McKenzie	Volunteer	ME	Newport Fire Dept
1/18/90	Gary M. Passaro	Volunteer	CT	Tolland Fire Dept
1/26/90	John W. Folds	Volunteer	GA	Heard County Emergency Services
1/28/90	Robert T. Crutchfield, III	Career	VA	Danville Fire Dept
1/28/90	Vernon D. Deshazor	Career	VA	Danville Fire Dept
1/28/90	Willard C. Kuhn	Unknown	PA	Goodwill Fire Co #1
1/29/90	Ronald L. Stroud	Career	OK	Nichols Hills Fire Dept
2/19/90	Clayton M. Cutter	Career	CA	USDA Forest Service
2/19/90	Ingrid H. Sowle	Volunteer	NY	Port Washington Fire Dept
2/19/90	Vidar D. Anderson	Volunteer	CA	Long Valley Fire Dept
2/20/90	Richard A. Havens	Volunteer	MI	A1mont Fire Dept
2/20/90	Richard V. Brekrus	Career	IN	South Bend Fire Dept
2/26/90	Dale M. Seib	Volunteer	NY	Plainville Fire Dept
3/10/90	Roger A. Houghton	Volunteer	MA	Templeton Fire Dept
3/11/90	Robert E. Lee	Volunteer	SC	Cowpens Fire Dept
3/13/90	Robert L. Hitchcock	Volunteer	NY	Blenheirn Hose Co
3/18/90	Floyd M. Price	Volunteer	VA	Luray Vol Fire Dept
3/18/90	John L. Kelley, Jr.	Volunteer	NY	Canisteo Vol Fire Dept
3/26/90	James D. Straub	Volunteer	MO	Western Taney County Fire Protection Dist
3/30/90	William P. Grimes	Volunteer	WV	Cool Springs Vol Fire Dept
4/6/90	Nicholas W. Hart	Volunteer	IN	Moorefield Community Fire Dept #1
4/8/90	Edward P. Dougherty, Jr.	Volunteer	OK	Fallis Rural Fire Fighters Association
4/19/90	Karl J. Drews	Career	FL	San Carlos Park Fire Dist
4/19/90	Kaye F. Anderson	Career	FL	Upper Captiva Vol Fire Dept
4/21/90	Anthony L. Boyert	Career	SC	St. Paul's District Fire Dept
4/22/90	Curtis C. Thomason	Volunteer	GA	Cornelia Fire Dept
5/1/90	James E. Chestnut, Jr.	Volunteer	PA	Hustontown Area Vol Fire Co
5/1/90	Richard L. Hershey	Volunteer	PA	Hustontown Area Vol Fire Co
5/1/90	Thomas L. Lane	Volunteer	PA	Hustontown Area Vol Fire Co
5/7/90	Mathe A. Alexander	Volunteer	MS	Long Creek Vol Fire Dept
5/10/90	Heriberto T. Rivera	Career	CT	Waterbury Fire Dept
5/10/90	Howard A. Hughes	Career	CT	Waterbury Fire Dept

Date of Incident	Full Name	Affiliation	State	Department
5/12/90	William McAdams	Volunteer	OR	Aurora Rural Fire Protection Dist
5/13/90	Mark A. Wunch	Volunteer	PA	Lake Shore Vol Fire Dept
5/15/90	James R. Ray	Career	AL	Montgomery Fire Dept
5/17/90	Robert L. Adams, Sr.	Volunteer	PA	Cresson Vol Fire Dept
5/17/90	Thomas N. Bianconi	Volunteer	PA	Cresson Vol Fire Dept
5/23/90	Hector M. Segura	Career	MA	Haverhill Fire Dept
5/28/90	Jerry A. Reed	Volunteer	MN	Deer River Fire Dept
6/1/90	Daniel R. Joslyn	Volunteer	NY	Maple Springs Vol Fire Dept
6/13/90	Frederick J. Heimann, Jr.	Volunteer	PA	Neffs Fire Co
6/14/90	Karl Richter	Volunteer	NY	Wantagh Fire Dept
6/17/90	Curtis McClain	Volunteer	WV	Ravenswood Aluminum Co. Security & Fire Protection
6/17/90	Peter Baltic	Volunteer	WV	Ravenswood Aluminum Co. Security & Fire Protection
6/19/90	James Goode, Jr.	Career	DE	Wilmington Fire Dept
6/19/90	Thomas D. Brashears	Career	KY	Hopkinsville Fire Dept
6/20/90	Ken M. Herington	Volunteer	IL	Washburn Fire Protection Dist
6/26/90	Alex S. Contreras	Volunteer	AZ	Arizona State Land Dept
6/26/90	Curtis E. Springfield	Volunteer	AZ	Arizona State Land Dept
6/26/90	James E. Ellis	Volunteer	AZ	Arizona State Land Dept
6/26/90	James L. Denney	Volunteer	AZ	Arizona State Land Dept
6/26/90	Joseph L. Chacon	Volunteer	AZ	Arizona State Land Dept
6/26/90	Sandra J. Bachman	Volunteer	AZ	Arizona State Land Dept
6/27/90	Aaron J. Perry	Volunteer	CA	California Dept of Forestry & Fire Protection
6/27/90	Victor Ferrera	Volunteer	CA	California Dept of Forestry & Fire Protection
6/30/90	Michael E. Van Calbergh	Volunteer	MI	Gibraltar Fire Dept
7/9/90	Daniel J. Raskin	Volunteer	MD	Chestnut Ridge Vol Fire Co
7/10/90	Fred R. Garber	Volunteer	PA	New Danville Fire Co
7/14/90	Edwin L. Simpson	Volunteer	NY	Mellenville Vol Fire Dept
7/17/90	George J. Karl	Volunteer	NY	Main Transit Fire Dept
7/27/90	John J. Meyer	Volunteer	OH	Craig Beach Village Fire Dept
7/28/90	Leonard E. Mills, Sr.	Career	TX	Midland Fire Dept
7/30/90	Mark A. Moore	Career	CA	USDA Fire Service
8/6/90	Toni J. Godsil	Volunteer	WA	Grand Coulee Vol Fire Dept
8/13/90	Kenneth E. Enslow	Career	CA	California Dept of Forestry & Fire Protection
8/15/90	Julius C. Starr	Career	OR	USDA Forest Service
8/21/90	Rufus J. Harrison	Volunteer	FL	Wetumpka Vol Fire Dept

Date of Incident	Full Name	Affiliation	State	Department
8/22/90	Mark E. Polan	Volunteer	VA	Blacksburg Vol Fire Dept
8/24/90	Barry M. Gray	Volunteer	KY	East Golden Pond Vol Fire Dept
8/25/90	Thomas Weeks	Volunteer	MD	Level Vol Fire Co
8/26/90	Blake V. Wright	Career	UT	Utah State Lands and Forestry
8/26/90	Ralph M. Broadhead	Career	UT	Utah State Lands and Forestry
9/6/90	Todd D. Colton	Career	KS	Sedgewick County Fire Dept
9/11/90	Robert P. Wiebe	Volunteer	WA	Colville Indian Agency, Contractor to BIA
9/23/90	Paul E. Caywood, Sr.	Volunteer	NY	Fabius Fire Dept
9/25/90	Robert J. Ely	Volunteer	IL	Holbrook Fire Dept
9/30/90	Ralph F. Glasgow	Volunteer	WA	USDA Forest Service -Contractor
9/30/90	Stephen E. Bovey	Volunteer	WA	USDA Forest Service -Contractor
10/11/90	George S. Labance	Volunteer	NJ	Franklin Fire Dept
10/12/90	Steve L. Giradot	Volunteer	GA	Damascus Rural Vol Fire Dept
10/19/90	Ronnie E. Hoots	Volunteer	NC	Edneyville Fire and Rescue Dept
10/20/90	Charles R. Lowery, II	Volunteer	NC	Lowell Vol Fire Dept
10/24/90	Lydia A. Sexton	Volunteer	TX	Spillway Fire Dept
10/27/90	Bruce A. Rhinehart	Career	NY	Jamestown Fire Dept
11/1/90	Lance J. Petersen	Career	CA	Oakland Fire Dept
11/5/90	Richard A. Penning	Unknown	WI	Kohler Fire Dept
11/11/90	Bennie B. Collins	Career	CA	Los Angeles City Fire Dept
11/13/90	Timothy M. Stine	Volunteer	PA	Lebanon City Bureau of Fire
11/14/90	Ronnie M. McAndrew	Career	TX	Arlington Fire Dept
11/18/90	Gene K. Copple	Volunteer	NE	Homer Vol Fire Dept
11/30/90	William F. Carter, Sr.	Volunteer	NY	Alcan Rolled Products Co. Fire Brigade
12/10/90	Robbie Davis	Volunteer	TN	Newmansville Vol Fire Dept
12/11/90	Robert J. Drennan	Volunteer	MS	Collins Vol Fire Dept
12/16/90	Thomas E. Hicks	Volunteer	PA	Citizens Vol Fire Co
12/20/90	James C. Winters	Career	FL	Spring Hill Fire and Rescue Dist
12/21/90	Edgar Dear	Volunteer	TX	Somerset Vol Fire Dept
12/22/90	Ralph L. Hoppenjans	Volunteer	IN	Ferdinand Vol Fire Dept
12/23/90	Thomas H. Maxeiner	Career	IL	Decatur Fire Dept
12/24/90	Loren N. Christian	Volunteer	WA	Chewelah Fire Dept
12/27/90	Leon L. Benton	Career	FL	Jacksonville Fire Dept
12/30/90	Ray J. Shenefield	Volunteer	KY	John's Creek Vol Fire Dept
1/1/91	Frank A. Quadrel	Career	NJ	Montclair Fire Dept
1/9/91	James E. Howe	Career	CA	Los Angeles County Fire Dept
1/10/91	Charles Love	Volunteer	MI	Boardman Twp Fire Dept
1/10/91	Edward W. Murdock, Sr.	Volunteer	NY	Port Ewen Fire Dept

Date of Incident	Full Name	Affiliation	State	Department
1/12/91	Curtis D. Mikkelsen	Volunteer	SD	Hurley Vol Fire Dept
1/17/91	George Hollowniczky	Volunteer	PA	Summer-Hill Twp Vol Fire Dept
1/17/91	John A. Nicosia	Career	NJ	Paterson Dept of Public Safety
1/19/91	Ronald C. Altieri	Career	CT	Hamden Fire Dept
1/20/91	Lynn E. Maricle	Volunteer	VA	Front Royal Vol Fire Dept
1/25/91	Robert Henry	Volunteer	MD	Trappe Vol Fire Dept
1/26/91	Edward E. Soper	Volunteer	PA	Leraysville-Pike Vol Fire Dept
1/27/91	John N. Plummer	Career	MD	Baltimore City Fire Dept
1/28/91	Brian T. Dillon	Career	NY	Buffalo Fire Dept
2/15/91	Robert F. Million	Volunteer	AZ	Avra Valley Fire Dist
2/15/91	William H. Marshall, Jr.	Volunteer	NY	Roosevelt Fire Dist
2/16/91	Joseph J. Alfred	Career	TX	Galveston Fire Dept
2/23/91	Wylie J. Morvant	Volunteer	LA	Coteau Vol Fire Dept
2/24/91	David F. Holcombe	Career	PA	Philadelphia Fire Dept
2/24/91	James A. Chappell	Career	PA	Philadelphia Fire Dept
2/24/91	Phyllis McAllister	Career	PA	Philadelphia Fire Dept
3/1/91	Marshal E. Viloria	Volunteer	CA	Newberry Springs Vol Fire Dept
3/5/91	Alfred E. Ronaldson	Career	NY	New York City Fire Dept
3/7/91	James E. Dame	Career	WY	Natrona County Fire Dept
3/8/91	Randall P. McDonald	Volunteer	MO	Caruthersville Fire Dept
3/12/91	Donald V. Mello	Career	CA	Los Angeles City Fire Dept
3/13/91	Eugene Miller	Volunteer	NY	Port Chester Fire Dept
3/13/91	Jennifer L. Kibbey	Volunteer	OH	Hilliar Twp -Centerburg Vol Fire Dept
3/15/91	Daniel E. Wannagot	Volunteer	CT	Shelton Fire Dept
3/15/91	Lou Falconer	Volunteer	CA	Boulevard Fire & Rescue Dept
3/19/91	Joseph D. Farrell	Volunteer	NY	Bayport Fire Dept
3/24/91	Donald I. Daughenbaugh	Volunteer	MI	Romulus Fire Dept
3/26/91	Frank L. Butler	Volunteer	KS	Marysville Vol Fire Dept
3/28/91	Alston F. Hill, Sr.	Volunteer	NY	Jamison Road Fire Dept
3/30/91	Robert I. Parker	Volunteer	VT	East Montpelier Vol Fire Dept
4/4/91	Brooks Cowgill	Volunteer	OH	Plain Twp Fire Dept
4/13/91	L. Wayne Struble	Volunteer	AZ	Williamson Valley Vol Fire Dept
4/14/91	John H. Spencer	Career	IL	Quincy Fire Dept
4/15/91	Timothy P. Scarborough	Volunteer	WV	Amstrong Creek Vol Fire Dept
5/28/91	Michael Moriarty	Volunteer	CT	Canterbury Vol Fire Co
5/31/91	Scott T. Laverty	Volunteer	NY	Hempstead Fire Dept
5/31/91	William F. Martin	Career	ID	USDA Forest Service, Nezperce National Forest
6/1/91	Eddie Arthur	Career	MD	Baltimore City Fire Dept

Date of Incident	Full Name	Affiliation	State	Department
6/9/91	Alfornia Hollis	Career	GA	Atlanta Fire Dept
6/9/91	Charles H. Boyle	Volunteer	PA	Greenwood hose Co #1
6/19/91	Daniel G. Paris	Volunteer	PA	Oscoluwa Engine and Hose Co
6/21/91	Henry Young Hi Kim	Career	NM	USDA Forest Service
6/21/91	Randolph F. Belcher	Volunteer	NY	Horseheads Fire Dept
6/22/91	Steven E. Bryant	Volunteer	KY	Feds Creek Fire Dept
6/24/91	Albert F. Robibero	Volunteer	NJ	Stanhope Fire Dept
6/25/91	James Hugh Lee	Volunteer	NC	Wilson's Mills Vol Fire Dept
6/29/91	Robert D. Morris, II	Career	SC	Charleston City Fire Dept
7/2/91	Marshall F. Sarles	Volunteer	NY	Mamaroneck Village Fire Dept
7/5/91	James T. Swindle	Volunteer	TN	Sumner County Vol Fire Dept
7/6/91	Wayne Ronald Russell	Career	FL	Orange County Fire Dept.
7/7/91	John G. Dugan	Volunteer	PA	Jim Thorpe Fire Dept
7/11/91	Joseph P. Kail	Volunteer	MI	New Buffalo City Fire Dept
7/16/91	Russell E. Dunham	Volunteer	NY	Highland Hose Co #1, Inc
7/26/91	Gifford Keeth	Career	CA	USDA Forest Service, Stanislaus National Forest
7/28/91	Raymond L. Bryant	Volunteer	MS	Bailey Vol Fire Dept
8/8/91	Jerry Brooks	Volunteer	SC	North Greenville Fire Dept
8/19/91	Robert J. Lucier, Sr.	Volunteer	NY	Alplaus Vol Fire Dept
8/21/91	Wayne L. Pulley	Career	CA	Ventura County Fire Dept
9/2/91	James D. Sapp	Career	FL	New Smyrna Beach Fire Dept
9/2/91	Mark A. Wilkes	Career	FL	New Smyrna Beach Fire Dept
9/7/91	Johnny Lewis, Jr.	Career	LA	New Orleans Fire Dept
9/12/91	Kevin C. Kane	Career	NY	New York City Fire Dept
9/15/91	Craig W. Keith	Volunteer	ND	Sherwood Fire Dept
9/16/91	Raymond Hazuka	Volunteer	KS	Wakeeney City Fire Dept
9/18/91	Kevin D. Keel	Career	CO	USDA Forest Service, Routt National Forest
9/28/91	Arthur R. Denny, II	Volunteer	OH	Franklin Twp Fire Dept
10/7/91	Nonnan L. Simmons	Career	GA	Savannah Fire Dept
10/11/91	Nathan E. Walls	Volunteer	MS	Calhoun City Vol Fire Dept
10/14/91	Frederick W. Templin	Career	MN	Saint Cloud Fire Dept
10/16/91	John R. Sieglinger	Career	CA	USDA Forest Service R5, Los Padres National Forest
10/16/91	Robert A. Shaw	Career	CA	USDA Forest Service R5, Los Padres National Forest
10/18/91	George S. Winckler	Career	IL	Naperville Fire Dept
10/20/91	James M. Riley, Jr.	Career	CA	Oakland Fire Dept
10/21/91	Joseph R. Bow	Volunteer	NY	Pearl River Fire Dept

Date of Incident	Full Name	Affiliation	State	Department
10/22/91	John A. Du Chateau	Volunteer	WI	Luxemburg Fire Dept
10/29/91	John E. Spangler	Volunteer	KY	Neon Vol Fire Dept
10/29/91	John R. Adams	Volunteer	KY	Neon Vol Fire Dept
10/29/91	Robert H. Foster, Sr.	Career	WV	Wheeling Fire Dept
11/4/91	Gregory E. Williams	Volunteer	NY	Holland Vol Fire Dept
11/5/91	Francis E. Nason	Volunteer	ME	Saco Fire Dept
11/6/91	John A. Kucich	Career	NJ	West New York Fire Dept
11/11/91	Gordon O. Schmidt	Volunteer	NY	St. Paul Boulevard Fire Dept
11/21/91	Clyde A. Burley	Volunteer	MD	Corriganville Vol Fire Dept
11/24/91	Joseph P. McCarthy	Career	NJ	Newark Fire Dept
11/29/91	John R. Cochran	Volunteer	AL	Mount Vera Vol Fire Dept
12/1/91	Mark S. Rice	Volunteer	IL	South Pekin Vol Fire Dept
12/3/91	Albert R. Smith	Career	PA	Pittsburgh Fire Dept
12/6/91	Daniel J. Miller, Jr.	Volunteer	NY	Getzville Fire Co
12/8/91	John H. Sienknecht	Volunteer	IA	Clutier Fire Dept
12/9/91	James D. Walling	Volunteer	MS	Woolmarket Vol Fire Dept
12/13/91	George R. Whiteside	Career	TX	Odessa Fire Dept
12/13/91	Kinnison F. Cribley	Volunteer	OH	Oregon Fire Dept
12/16/91	Fred P. Biedron	Volunteer	IN	Hammond Fire Dept
12/18/91	Thomas E. Coyne, Jr.	Career	OH	Brooklyn Fire Dept
12/19/91	Robert A. Cole	Volunteer	CT	Beacon Hose Co #1
12/20/91	David G. Emanuelson	Volunteer	PA	Hilltop Hose Co #3
12/20/91	Frank Veri, Jr.	Volunteer	PA	Hilltop Hose Co #3
12/20/91	Michael J. Cielicki	Volunteer	PA	Hilltop Hose Co #3
12/20/91	Richard A. Frantz	Volunteer	PA	Hilltop Hose Co #3
12/20/91	Stephen D. Yale	Career	PA	Philadelphia Fire Dept
12/23/91	Larry H. Brooks	Career	VA	Portsmouth Fire Dept
1/5/92	James W. Schott	Volunteer	PA	Good Year Hose Co
1/12/92	Kenneth M. Hedrick	Volunteer	MD	Prince George's County Fire Dept
1/16/92	Robert G. Reits, Sr	Volunteer	MI	Paw Paw Fire Dept
1/18/92	Raymond E. Talley	Volunteer	PA	Concordville Fire Co
1/21/92	James A. Bennett	Career	AL	Birmingham Fire Dept
1/29/92	Roberto Valdez	Career	TX	Laredo Fire Dept
2/5/92	Elwood M. Gelenius	Career	IN	Indianapolis Fire Dept
2/5/92	John J. Lorenzano	Career	IN	Indianapolis Fire Dept
2/23/92	Daniel L. Maturen	Career	MI	Battle Creek Fire Dept.
2/24/92	Thomas A. Williams	Career	NY	New York City Fire Dept
2/28/92	Darrell R. Nam	Career	HI	Honolulu Fire Dept

Date of Incident	Full Name	Affiliation	State	Department
3/1/92	Gordon A. Champney	Volunteer	VT	Bolton Vol Fire Dept
3/2/92	Herbert B. Campbell	Volunteer	MD	Middle River Vol Fire Co
3/4/92	Clifford L. Hennan	Career	ID	USDA Forest Service
3/13/92	Arthur K. Tuck	Career	NY	New York City Fire Dept
3/15/92	Nelson Margerum	Volunteer	PA	Yardley-Makefield Fire Co
3/23/92	Billy Ray Powell	Volunteer	AR	Lepanto Vol Fire Dept
4/4/92	Richard C. Heller, Jr.	Volunteer	IL	Elk Grove Twp Fire Dept
4/21/92	Roy J. Swinehart	Volunteer	OH	Carroll County Vol Fire Dept
5/18/92	George William Ousley	Volunteer	OK	Cowskin Fire Dept
5/26/92	Frank Albert Smith	Career	FL	Cape Coral Fire Dept
6/8/92	Wayne E. Walker	Career	KS	Kansas City Fire Dept
6/14/92	Roland Waters	Career	MI	Detroit Fire Dept
6/16/92	Thomas I. Horvath	Career	PA	Lancaster Bureau of Fire
6/19/92	Guy C. Miller	Volunteer	PA	Light Street Community Fire Co
6/19/92	Roger Stark	Contract	CA	California Dept of Forestry and Fire Protection
6/20/92	Robert S. Heide, Sr.	Volunteer	CT	Plainfield Fire Co # 1
6/29/92	Richard E. Wilson	Volunteer	CO	Glenwood Springs Dept of Emergency Services
6/30/92	James H. Stavely, III	Volunteer	MD	Braddock Heights Vol Fire Dept
7/4/92	Paul K. Bjorkland	Volunteer	WA	Benton County Fire Protection Dist #1
7/12/92	Bryan Lo Weeks	Career	MO	Potosi Fire Dept
7/15/92	Kim Meredith	Volunteer	IL	Sangamon Valley Fire Protection Dist
7/27/92	Lupe Guerrero	Volunteer	TX	Menard Vol Fire Dept
7/29/92	Patrick J. Luby	Career	IL	Oak Park Fire Dept
7/29/92	Robert Medlicott	Career	IL	Berwyn Fire Dept
7/30/92	Julie Ann Young	Seasonal	ID	USDA Forest Service
7/31/92	Dean O. Pucker	Volunteer	WI	Lamartine Vol Fire Dept
8/3/92	Lionel Alves	Career	MA	New Bedford Fire Dept
8/9/92	David K. Lumbra	Volunteer	NH	Charlestown Fire Dept
8/16/92	Calvin Morris	Volunteer	AZ	Mayer Fire Dist
8/25/92	John M. Byers	Career	FL	Ocean Reef Vol Fire Dept
8/28/92	Corey R. Clawson	Volunteer	MT	Libby Vol Fire Dept
8/31/92	James Shannon Campbell	Contract	OR	Oregon Dept of Forestry
9/1/92	Heather L. Dzioba	Career	WI	Milwaukee Fire Dept
9/5/92	Charles E. Fierson	Volunteer	PA	Stroudsburg Fire Dept
9/9/92	Eldon W. Harrison	Volunteer	IL	Depue Fire Dept
9/20/92	Earle V. Dudley, III	Volunteer	NH	Middleton Vol Fire Dept
9/28/92	Mark W. Langvardt	Career	CO	Denver Fire Dept

Date of Incident	Full Name	Affiliation	State	Department
9/30/92	Robert D. Barnes	Volunteer	PA	Minersville Rescue Hook & Ladder Co
10/1/92	Charles F. Sheridan	Contract	CA	USDA Forest Service
10/1/92	Chester F. Warner	Volunteer	RI	Prudence Island Fire Dept
10/1/92	Leonard D. Martin	Contract	CA	USDA Forest Service
10/12/92	Anthony Vo Calhoun	Volunteer	LA	DeSoto Fire Dist #8
10/14/92	Michael D. Hoover	Volunteer	NC	Tabernacle Fire Dept
10/16/92	Victor E. Eager	Volunteer	PA	Brecknock Twp Fire Co #1
10/20/92	Gwyn L. Ellis	Volunteer	VA	LaCrosse Vol Fire Dept
10/24/92	Gary S. Porter	Career	IL	Alton Fire Dept
10/24/92	Tim L. Lewis	Career	IL	Alton Fire Dept
10/25/92	Arthur Lo Pearson	Volunteer	KS	Ogden Fire Dept
10/29/92	John A. Navarro	Volunteer	MI	Sandusky Fire Dept
10/30/92	Harold J. Lyons, Sr.	Volunteer	NY	Brookhaven Fire Dept
11/1/92	Anthony J. Carugno	Volunteer	NJ	Egg Harbor City Fire Dept
11/4/92	Tommy A. Parker	Volunteer	TX	Mineral Wells Fire Dept
11/6/92	Roc E. Manchester	Career	CA	Fire Dept MCAS El Toro
11/8/92	Wilbert L. Dyer	Volunteer	KS	Jefferson County Fire Dist 11
11/17/92	Don L. Milner	Volunteer	IA	Elliott Fire Dept
11/22/92	William N. Jones, Jr.	Volunteer	SC	Iva Fire Dept
12/1/92	Rebecca Ann LeClaire	Volunteer	TX	Brazos County Precinct #4 Fire Dept
12/5/92	Jeffrey G. Osmun	Volunteer	TX	Crowley Fire Dept
12/17/92	James J. Cothran	Volunteer	AL	Pine Level Vol Fire Dept
12/26/92	James D. Hill	Career	TN	Memphis Fire Dept
12/26/92	Jospeh A. Boswell	Career	TN	Memphis Fire Dept
12/28/92	George Delbert Weischman	Career	AR	Stuttgart Fire Dept
12/31/92	Arthur E. Schumacher	Volunteer	OH	Valley City-Liverpool Fire Dept
1/2/93	Cecil A. Fain	Volunteer	KY	Jessamine County Fire Dist
1/2/93	Harry B. Garis	Volunteer	PA	Logan Fire Co 1
1/2/93	William M. Wheeler	Volunteer	KY	Jessamine County Fire Dist
1/4/93	Ronald L. O'Rouke	Volunteer	NY	Roosevelt Dist Fire Dept
1/8/93	Walter E. Siverton, Jr.	Career	VA	Newport News Fire Dept
1/9/93	Norman G. Schunk, Jr.	Volunteer	NY	East Eden Fire Co.
1/10/93	Richard A. Hartley	Volunteer	PA	East Derry Volunteer Fire Dept
1/16/93	Joseph A. Hummel	Volunteer	PA	Friedensburg Fire Dept
1/23/93	Francis M. Nichols	Volunteer	NE	Imperial Fire Dept
1/26/93	Tony F. Mendonsa	Volunteer	CA	Turlock Rural Fire Dept
1/29/93	Howard R. Schmitt	Career	NY	Albany Fire Dept
1/29/93	Peter S. Sadowski	Volunteer	MA	Webster Fire Dept

Date of Incident	Full Name	Affiliation	State	Department
1/31/93	Douglas K. Konecny	Career	CO	Denver Fire Dept
1/31/93	William Grounds	Volunteer	TX	Neches Volunteer Fire Dept
2/1/93	John M. O'Conner	Career	CT	New London Fire Dept
2/5/93	Rick A. Vreeland	Career	FL	Gray Gables Nassau Village Vol Fire Dept
2/18/93	Richard C. McQuaide, Sr.	Volunteer	TX	Granite Shoals Fire Dept
2/24/93	Warren R. Ogbum	Career	NY	White Plains Fire Dept
2/27/93	Brian E. Metts	Volunteer	OH	North Georgetown Fire Dept
2/27/93	Christopher P. Savage	Volunteer	NY	Sir William Johnson Fire Dept
3/6/93	Dennis R. Olson	Career	IL	Monmouth Fire Dept
3/7/93	Keith A. Walker	Volunteer	FL	Gray Gables Nassau Village Vol Fire Dept
3/8/93	Donald C. Cottrell	Career	OH	Toronto Fire Dept
3/10/93	George Blanusa	Volunteer	CT	Branford Fire Dept
3/13/93	Dennis Rodd	Volunteer	NY	Coram Fire Dept
3/15/93	John F. Lombardo	Volunteer	PA	Pittston City Fire Dept
3/15/93	Leonard C. Insalaco, II	Volunteer	PA	West Pittston Hose Co
3/15/93	Samual Isaac Boyce	Volunteer	VA	Shawnee Fire Co
3/19/93	Troy V. Henderson	Volunteer	WV	Frost Vol Fire Dept
3/20/93	Carlos A. Negron	Career	NJ	Jersey City Fire Dept
3/30/93	Jeffrey Langley	Career	CA	Los Angeles County Fire Dept
4/1/93	Lewis R. Sheats	Both	NJ	Chatham Boro Fire Dept
4/7/93	Patrick J. Dougherty	Volunteer	NC	Gamer Vol Fire Dept
4/10/93	Gary L. Kennicutt	Volunteer	NY	Greene Fire Dept
4/15/93	Loren E. Baker	Volunteer	NH	Holdemess Fire Dept
4/22/93	Frankie Toledo	Seasonal	NM	USDA Forest Service
4/24/93	William M. Overman, Jr.	Volunteer	MD	Anne Arundel County Fire Dept
4/30/93	Raymond Adkins	Volunteer	KY	Big Creek Vol Fire Dept
5/21/93	Jacob Cogdil Parris, Jr.	Volunteer	NC	Balsam-Willets-Ochre Hill VFD
5/24/93	Michael J. Wilcom, Jr.	Volunteer	MD	New Market Dist Fire Dept
6/20/93	David M. Dunkerley	Volunteer	VA	Louisa Vol Fire Dept
6/29/93	Edwin R. Conklin	Volunteer	IL	Earlville Fire Protection Dist
7/3/93	Randy W. Reynaga	Volunteer	CA	Fountain Valley Fire Dept
7/6/93	Richard Beck	Volunteer	NY	Owego Fire Dept
7/9/93	Richard E. Gleason, Sr.	Volunteer	NY	Lake Mohegan Dist Fire Dept
7/14/93	Allan F. Coates	Volunteer	OH	Lafayette-Jackson Twp Fire Dept
7/27/93	Delmar M. Mondy	Volunteer	IL	Seneca Vol Fire Dept
8/5/93	Larry R. Harris, Jr.	Volunteer	GA	Morgan County Vol Fire Dept
8/5/93	Louis T. Powers	Volunteer	GA	Morgan County Vol Fire Dept
8/15/93	Everett C. Pierce	Volunteer	FL	Georges Lake Area Fire Dept

Date of Incident	Full Name	Affiliation	State	Department
8/20/93	Arthur Ruezga	Career	CA	Los Angeles County Fire Dept
8/20/93	Christopher D. Herman	Career	CA	Los Angeles County Fire Dept
8/22/93	John F. Hargreaves	Career	RI	Pawtucket Fire Dept
8/24/93	James R. Lafon	Volunteer	SC	Horry County Fire Dept
8/25/93	Joseph D. Delvecchio, Sr.	Career	PA	Dunmore Fire Dept
8/27/93	Joseph E. Kozlowski	Federal	PA	Tobyhanna Army Depot
8/28/93	Arthur Karr	Volunteer	PA	Shohola Vol Fire Dept
9/6/93	Brian L. Hill	Seasonal	OR	Department of Forestry
9/18/93	Jimmy L. Jackson	Volunteer	TX	Fort Stockton Fire Dept
10/12/93	Francis J. Baker	Career	MA	Boston Fire Dept
10/13/93	Patrick Lafferty	Volunteer	NY	Greenville Vol Fire Dept
10/19/93	Richard H. Melloni, Sr.	Career	MA	Wareham Fire Dept
10/22/93	Christopher C. Rutledge	Career	CA	Lucerne Valley Fire Dept
11/11/93	Wilbert I. Hansen	Volunteer	WY	Laramie County Fire Dept
11/15/93	Gary W. Armstrong	Volunteer	KY	Mount Washington Fire Dept
11/15/93	Harold B. Allgood	Volunteer	KY	Mount Washington Fire Dept
11/15/93	John H. Somay	Volunteer	NJ	Southard Fire Dept
11/20/93	Charles H. Beadle	Volunteer	NY	Walton Fire Department
11/23/93	Steven J. McNamee	Career	IL	Chicago Fire Dept
11/30/93	Guy E. Post	Volunteer	KS	Bourbon County #3 Fire Dept
12/6/93	Dale E. Linkroum	Volunteer	NY	Deposit Fire Dept
12/6/93	Jesse Pacheco, Jr.	Career	MA	New Bedford Fire Dept
12/14/93	Russell T. Newcomb	Volunteer	NJ	South Amboy Fire Dept
12/15/93	Mark R. Hinson, Sr.	Volunteer	TN	Pairs Landing Vol
12/23/93	John Brentzel, Jr.	Volunteer	NJ	Carlstadt Fire Dept
12/23/93	Vincent D. Meegan, Jr.	Career	NY	Buffalo Fire Dept

Fatality and Incidence Summary: 1994–2000

1/1/1994   **Ronnie Fuller, Firefighter**   **Career, Age 49**
     Clinton Fire Department, SC   **Heart Attack**

Firefighter Fuller died after suffering a heart attack while operating a pumper at a house fire. His fellow firefighters and police officers on the scene immediately rushed to his aid when they saw him collapse and initiated CPR. He was transported to the hospital where he died.

1/3/1994   **Marcus Carr, Firefighter/Paramedic**   **Career, Age 54**
     David Mosher, Firefighter/Paramedic   **Career, Age 35**
     Chillicothe Fire Department, MO   **Trauma**

Firefighter Carr and Firefighter/Paramedic Mosher were killed when a tractor-trailer truck that veered into their lane struck their ambulance head on. The two were en route to the hospital with a patient, who also died in the crash. The ambulance was running with lights and siren at the time of the accident. The truck driver was critically injured.

1/4/1994   **Thomas Dunn, Firefighter**   **Volunteer, Age 40**
     Rutherford Fire Department, NJ   **Burns/Asphyxiation**

Firefighter Dunn was killed when he became trapped on the second floor by a rapidly advancing fire in a balloon frame house. Firefighter Dunn's company had been conducting search operations and horizontal ventilation in the house with only light fire and smoke conditions. The fire had initially been located in the basement. Conditions rapidly deteriorated as the pressurized heat and smoke broke out of concealed spaces in the attic above the fire fighters and from a dropped tin ceiling on the floor below them. Two personnel with Firefighter Dunn were able to escape through second floor windows as evacuation signals were sounded. Firefighter Dunn became disoriented and entangled in a bed frame on the second floor and was unable to escape. A rescue team entered the second floor via a window and quickly reached him, but he died from acute smoke inhalation and burns. His facepiece was not on when he was found, his SCBA straps had failed, and his PASS device was in the off position. A New Jersey Division of Fire Safety report indicates that the balloon frame construction, ventilation simultaneously below and above hidden fire areas, and operation of fog pattern streams in the basement may have all contributed to the rapidly deteriorating conditions faced by the fire crews on the second floor.

1/8/1994   **Gerald Mullins, Firefighter**   **Career, Age 55**
     Binghamton Fire Department, NY   **Heart Attack**

Firefighter Mullins had just finished a shift assigned to an EMS unit and left the firehouse when he collapsed from a heart attack across the street. Medical care was quickly administered but efforts to revive him were unsuccessful.

1/10/1994   **Harold Salisbury, Deputy Chief**   **Volunteer, Age 54**
     East Greenwich Fire Department, RI   **Heart Attack**

Deputy Chief Salisbury was incident commander at a structure fire in a metal product manufacturing plant. During the incident Chief Salisbury collapsed from a heart attack. Efforts to revive him were unsuccessful, and he was pronounced dead after transport to the hospital.

1/12/1994   **Dennis Mullins, Jr., Firefighter**   **Affiliation and Age Unknown**
     Mount Vernon Fire Department, NY   **Heart Attack**

Firefighter Mullins suffered a heart attack at a fire. He died of complications from that heart attack in 1995.

1/21/1994   **Glen Thorn, Firefighter**   **Volunteer, Age 68**
     Sea Girt Fire Company #1, NJ   **Heart Attack**

Firefighter Thorn suffered a heart attack at the scene of a structure fire after arriving in his personal vehicle and preparing to perform support functions at the exterior of the structure. His collapse was witnessed by other fire department personnel who administered emergency care. He was transported to the hospital and died 19 days later.

| 1/22/1994 | **George Ciliberto, Captain** | **Career, Age 54** |
| | **Ocean City Fire Department, NJ** | **Heart Attack** |

Captain Ciliberto died of a heart attack in his sleep while on duty at the fire station. Captain Ciliberto had run several emergency calls during the evening.

| 1/23/1994 | **Maurice Wardell, Jr., Firefighter** | **Volunteer, Age 58** |
| | **Proctor Fire Department, VT** | **Heart Attack** |

Firefighter Wardell collapsed and died of an apparent heart attack after arriving on the scene of a working fire.

| 1/28/1994 | **Nick Charmello, Captain** | **Career, Age 53** |
| | **Kansas City Fire Department, MO** | **Heart Attack** |

Captain Charmello had just finished assisting in the extrication of a patient from an automobile accident when he collapsed from an apparent heart attack. Firefighters and rescue crews on the scene rushed to his aid. He was pronounced dead after being transported to the hospital.

| 1/28/1994 | **Walter Franks, Firefighter** | **Volunteer, Age 66** |
| | **Pine Hill Fire Department, NJ** | **Heart Attack** |

Firefighter Franks stayed at the fire hall to prepare coffee for firefighters out on a call. They returned to find him unconscious, having suffered a heart attack. Emergency care was initiated. Franks died several days later at the hospital.

1/28/1994	**Vencent Acey, Firefighter**	**Career, Age 42**
	John Redmond, Firefighter	**Career, Age 41**
	Philadelphia Fire Department, PA	**Trapped**

Firefighters Acey and Redmond died when they became trapped and overcome by smoke by a rapidly moving fire in the basement of a church. Several firefighters re-entered the church against orders to rescue the firefighters, and were able to pull one of them from the basement. Eight other firefighters were injured including several involved in the rescue efforts.

| 2/1/1994 | **Marilyn Williams, Firefighter** | **Volunteer, Age 55** |
| | **Keystone Volunteer Fire Department, OK** | **Heart Attack** |

Firefighter Williams suffered a heart attack and died while pulling hose at a mobile home fire.

| 2/5/1994 | **Robert English, Battalion Chief** | **Career, Age 57** |
| | **Detroit Fire Department, MI** | **Heart Attack** |

Battalion Chief English suffered a fatal heart attack while directing crews during the overhaul stages of an apartment fire. Children playing with matches are believed to have caused the fire.

| 2/7/1994 | **Newt Morgan, Firefighter** | **Volunteer, Age Unknown** |
| | **Poughkeepsie Volunteer Fire Department, AR** | **Heart Attack** |

Firefighter Morgan was driving a fire engine to a reported structure fire when he suffered a massive heart attack. The engine veered slowly off the road into a tree. Firefighter Morgan was found in cardiac arrest by firefighters responding behind him and is believed to have died before the accident occurred.

| 2/11/1994 | **Timothy Hale, Fire Engineer** | **Career, Age 29** |
| | **Phoenix Fire Department, AZ** | **Struck by Vehicle** |

Fire Engineer Hale was killed when he and his partner were struck by a vehicle while unloading the stretcher from the rear of their rescue unit during an EMS incident. Hale received severe traumatic injuries and died at a trauma center the following day. The driver of the vehicle was intoxicated.

2/20/1994   **Ann F. Sheppard, Firefighter**   Career, Age 26
          Venus Volunteer Fire Department, FL   Heart Attack

Firefighter Sheppard was participating in search-and-rescue training when she suffered a fatal heart attack.

2/26/1994   **Bedford Cash, District Ranger**   Career, Age Unknown
          United States Forest Service   Heart Attack

District Ranger Cash was conducting prescribed burning in the Tuskegee National Forest when he suffered a fatal heart attack.

2/27/1994   **Dennis Dearing, Jr., Firefighter**   Volunteer, Age 27
          Auburn Hills Fire Department, MI   Trapped

Firefighter Dearing died when the floor collapsed under him while conducting operations at a house fire. Firefighter Dearing and two others had entered the house through the kitchen with a hose line to try to reach a fire located in the basement that had been burning for approximately 40 minutes. The officer in charge of the attack crew ordered them to evacuate the house due to the spongy feeling of the floor as they approached the basement stairs, but the floor collapsed. The fire was ruled incendiary in nature, with a high fire load of combustibles in the basement contributing to the floor collapse.

3/2/1994   **Mark Mitchell, Firefighter**   Volunteer, Age 26
          Pawcatuck Fire Department, CT   Carbon Monoxide Poisoning

Firefighter Mitchell died of carbon monoxide poisoning after being separated from his crew while conducting search operations on the second floor of a single-family house. It is believed that there was a delay of over 1hour before the fire department was called. Mitchell and three other firefighters were attempting to rescue a victim reported on the second floor when a flash-over occurred, separating the crewmembers. Three firefighters escaped with injuries; Mitchell was found unconscious on the second floor and died later. His blood carboxyhemoglobin level was 24%.

3/5/1994   **Charles Butchee, Firefighter**   Volunteer, Age 54
          Warren Community Fire Department, OK   Heart Attack

Firefighter Butchee died of a heart attack after being exposed to smoke and heat at an outside controlled burn.

2/6/1994   **Walter Wade, Lieutenant**   Volunteer, Age 34
          Miami Township Fire Department, OH   Heart Attack

Lieutenant Wade died of a heart attack after completing a search for victims at a house fire. An autopsy revealed that Lt. Wade had an enlarged heart and a genetic heart abnormality.

3/22/1994   **Gary King, Firefighter**   Volunteer, Age Unknown
          Grundy County Rural Fire Protection District, MO Heart Attack

Firefighter King suffered a heart attack while operating at a two-alarm brush fire. He died after being transported to the hospital.

3/22/1994   **Dustin Mills, Firefighter**   Volunteer, Age 22
          Capron Fire Department, OK   Apparatus Rollover

Firefighter Mills died en route to a wildfire that burned over 5,000 acres. The brush truck he was riding on overturned as it drove over a smoke-obscured 15-foot embankment. He died of traumatic injuries at the scene; another firefighter received minor injuries.

3/29/1994   **John Drennan, Captain**   Career, Age 49
          James Young, Firefighter   Career, Age 31
          Christopher Seidenburg, Firefighter   Career, Age 25
          New York Fire Department, NY   Burns

Captain Drennan and Firefighters Young and Seidenburg were conducting a search when the hot air and toxic gases that collected in the stairwell erupted into flames as other fire crews forced entry into the first floor apartment where the fire originated. The fire exhibited characteristics of both a backdraft and a flashover. Firefighter Young, in the bottom position on the stairs, was burned and died at the scene. Firefighter Seidenberg and Captain Drennan were rescued by other firefighters. They were transported to a burn unit with third and fourth degree burns over 50 of their bodies. Firefighter Seidenburg died the next day. Firefighter Drennan passed away several weeks later. The fire cause was determined to be a plastic bag left by the residents on top of the stove of the second floor apartment.

| 4/2/1994 | **Joseph Jay Boothe, Firefighter**
Pea Ridge Volunteer Fire Department, AL | **Volunteer, Age 17**
Apparatus Rollover |

Firefighter Boothe was riding in the passenger seat of a 1971 Ford 1,200-gallon tanker truck en route to a brush fire that was threatening several homes and a church. Boothe was killed when the vehicle overturned heading into a sharp turn. The police report indicated that the vehicle was traveling at approximately 35 miles per hour in a 30 mph zone. The driver received minor injuries and reported that the brakes on the vehicle locked up heading into the curve. The vehicle had no seat belts.

| 4/3/1994 | **Robert Waskiewicz, Firefighter**
Augusta–Bridge Creek Fire Department, WI | **Volunteer, Age 31**
Burns |

Firefighter Waskiewicz received fatal burn injuries when he was caught in a wind shift and overrun by a fast moving grass fire.

| 4/7/1994 | **Ronald Carlson, Firefighter**
Blue Creek Rural Fire Protection District/
Lewellen Volunteer Fire Department, NE | **Volunteer, Age 55**
Apparatus Rollover |

Firefighter Carlson was driving fire apparatus to a brush fire when the vehicle rolled over, killing him and seriously injuring two other firefighters.

| 4/11/1994 | **Michael Mathis, Lieutenant**
William Bridges, Private
Memphis Fire Department, TN | **Career, Age 39**
Career, Age 27
Trapped |

Lieutenant Mathis and Private Bridges were killed when they became trapped and overcome by smoke during a fire on the ninth floor of a high-rise building. Two civilians also died in the arson fire. Lt. Mathis became disoriented when he was caught in rapidly spreading fire conditions on the fire floor, burning him and causing his SCBA to malfunction. He found his way into a room on the ninth floor where he was later discovered by other fire crews with his SCBA air depleted. Private Bridges, aware that Lt. Mathis was unaccounted for after several unsuccessful attempts to contact him by radio, left a safe stairwell where he had been attempting to fix problem with his own SCBA. Investigators believe Private Bridges was trying to locate Lt. Mathis, became entangled in fallen cable TV wiring within a few feet of the stairwell, and died of smoke inhalation after depleting his SCBA supply. A Memphis Fire Department investigation found many violations of standard operating procedures by companies on the scene, including crews taking the elevator to the fire floor, problems with the incident command system and coordination of companies, operating a ladder pipe with crews still on the fire floor, and a failure of personnel, including Lt. Mathis and Private Bridges, to activate their PASS devices.

| 4/15/1994 | **Stanley Rhoads, Firefighter**
Barrick Goldstrike Emergency Response Team, NV | **Career, Age 47**
Smoke Inhalation |

Firefighter Rhoads was on his way to work when a fire broke out in a gold refinery building. After arriving, he was witnessed putting on his personal protective clothing and SCBA. Two hours later, members of a volunteer fire department that had responded to the fire found his body inside the fire building. He had apparently entered the structure independently and ran out of air inside the refinery. Commanders did not know he was on the fire scene until his body was removed. The initial fire attack was described as "hectic" to the Nevada State Fire Marshal that investigated the report.

4/29/1994	Joseph Jarvis, Sr., Fire/Police Officer	Affiliation and Age Unknown
	Oceanside Fire Department, NY	Struck by Vehicle

Fire/Police Officer Jarvis was struck and killed by a vehicle while directing traffic at an emergency scene.

5/30/1994	Alton Warren, Firefighter	Career, Age 54
	Baltimore City Fire Department, MD	Embolism

Firefighter Warren was injured when he fell down the stairs at a fire, breaking his ankle. He died later of an embolism that developed from the injury.

6/4/1994	Anthony Covis, Senior Firefighter	Career, Age 42
	Newport News Fire Department, VA	Heart Attack

Senior Firefighter Covis died after suffering a heart attack while on duty in his station. Firefighter Covis, a 20-year veteran, had participated in a morning physical training exercise, and had gone to a separate room in the firehouse after eating lunch. Fellow firefighters found him in cardiac arrest a few hours later.

6/5/1994	George Lener, Lieutenant	Career, Age Unknown
	New York City Fire Department, NY	Smoke Inhalation

Lieutenant Lener collapsed from smoke inhalation and was found unconscious in the basement of a five story warehouse after a fire that required the response of more than 300 firefighters. Lt. Lener died 7 weeks later without regaining consciousness. A suspected arsonist has been arrested and charged with starting the fire. Sixteen other firefighters were injured during the incident.

6/10/1994	Victor Ruth III, Firefighter	Volunteer, Age 37
	Neptune Fire Company #1, PA	Heart Attack

Firefighter Ruth suffered a fatal heart attack while responding as part of an engine company to a medevac standby.

6/13/1994	Marc Butcher, Private	Affiliation Unknown, Age 37
	Parkersburg Fire Department, WV	Heart Attack

Private Butcher died in his sleep of a heart attack while on duty at the fire station. Efforts by fellow firefighters to revive him were unsuccessful.

6/14/1994	Ronald Holmgreen, Firefighter	Career, Age 46
	Lake Havasu Fire Department, AZ	Heart Attack

Firefighter Holmgreen suffered a fatal heart attack shortly after returning home from a fire department drill. Firefighter Holmgreen had exhibited signs of cardiac distress during the drill.

6/18/1994	David Barter, Firefighter/EMS Coordinator	Volunteer, Age 58
	West Terre Haute Volunteer Fire Department, IN	Heart Attack

Firefighter/EMS Coordinator Barter died after suffering a heart attack at his station. He had just returned from an emergency medical call in very hot weather.

6/24/1994	Stephen Minehan, Lieutenant	Career, Age 44
	Boston Fire Department, MA	Smoke Inhalation

Lieutenant Minehan died after leading his company in a successful search for two other trapped firefighters at a blaze in a vacant waterfront warehouse. He apparently became disoriented in the heavy smoke conditions and was separated from his company as they rescued the trapped firefighters. He radioed that he was trapped but several rescue efforts to find him were unsuccessful. He died of smoke inhalation and his company recovered his body several hours later.

6/28/1994	Clifford Harris, Chief	Volunteer, Age 50
	Rusk Volunteer Fire Department, TX	Heart Attack

Chief Harris responded to a house fire. After the fire was knocked down, Chief Harris entered the structure to assist with overhaul operations in the area of origin. Chief Harris was not wearing turnout gear or SCBA. After several minutes, Chief Harris left the structure and collapsed in cardiac arrest. He had had heart bypass surgery 10 years earlier. The fire had burned several hundred plastic videocassette tapes, which produced toxic gases. Chief Harris' death was attributed to a heart attack caused by inhalation of toxic gas; no autopsy was performed.

| 6/29/1994 | Anthony Bullard, Fire Marshal | Career, Age 42 |
| | Greenville Fire Department, TX | Cerebral Hemorrhage |

Fire Marshal Bullard suffered a cerebral hemorrhage during physical education class at the police academy. He died 6/30/1994.

7/6/1994	Don Mackey, Smokejumper	Age 34
	Roger Roth, Smokejumper	Age 31
	James Thrash, Smokejumper	Age 44
	Jon Kelso, Prineville Hot Shots	Age 27
	Kathi Beck, Prineville Hot Shots	Age 24
	Scott Blecha, Prineville Hot Shots	Age 27
	Levi Brinkley, Prineville Hot Shots	Age 22
	Bonnie Holtby, Prineville Hot Shots	Age 21
	Rob Johnson, Prineville Hot Shots	Age 26
	Terri Hagen, Prineville Hot Shots	Age 28
	Doug Dunbar, Prineville Hot Shots	Age 22
	Tami Bickett, Prineville Hot Shots	Age 25
	Robert Browning, Helitack	Age 28
	Richard Tyler, Helitack	Age 33
		Overrun by Wildfire

Fourteen wildland firefighters lost their lives when a wind shift resulted in a blow-up fire condition that trapped them on the uphill and downwind position from a fire on Storm King Mountain, Colorado. The 14 firefighters included smokejumpers Mackey, Roth, and Thrash; Prineville Hot Shots Kelso, Beck, Blecha, Brinkley, Holtby, Johnson, Bickett, Dunbar, and Hagen; and Helitack crewmembers Tyler and Browning. Browning and Tyler were killed when a large drop cut off their escape route and the fire overran them. The other firefighters were killed as they moved toward the ridgeline to escape the fire advancing toward them from below. According to witness accounts, the firefighters were unable to see how dangerous their position had become because of a small ridge below them. They had been moving slowly and were still carrying their equipment as the fire blew up behind them to a height of over 100 feet. At this point the crew dropped their tools and made an uphill dash for the top of the mountain but only one person made it over to survive. The fire overran the remaining 12 firefighters and reportedly reached a height of 200 to 300 feet as it crossed over the ridge. It was estimated to be moving at between 10 and 20 miles per hour at the time of the blow up. Several other firefighters in various other locations on the mountain became trapped by the flames but were able to make it to safe positions or deploy their emergency shelters. Post-incident investigations have determined that the crews fighting the fire violated many safety procedures and standard firefighting orders. The weather conditions prevalent that day had forecast a "red flag," the most dangerous wildfire conditions.

7/12/1994	Robert Boomer, Pilot	Career, Age Unknown
	Sean Gutierrez, Firefighter	Career, Age Unknown
	Sam Smith, Firefighter	Career, Age Unknown
	Briles Wings and Helicopter and Helitack	Helicopter Crash

Pilot Boomer and Helitack Firefighters Gutierrez and Smith were killed when their helicopter crashed while transporting them to a wildfire burning in the Black Range of the Gila National Forest. Two other crewmembers were injured in the crash.

| 7/23/1994 | Michael Shaughnessy, Firefighter | Career, Age 32 |
| | Cleveland Fire Department, OH | Trauma |

Firefighter Shaughnessy was killed when he fell off the roof of his fire station.

| 7/27/1994 | Paul Hodges, Firefighter | Seasonal Wildland, Age Unknown |
| | United States Forestry Service | Heart Attack |

Firefighter Hodges died after suffering a heart attack while driving a tanker truck at a wildland fire. He was a contract employee for the USFS and a volunteer with the Chelan County Fire Protection District #9.

7/29/1994	Robert Kelly, Pilot	Career, Age 58
	Randy Lynn, Pilot	Career, Age 44
	Neptune, Inc. (Contract to USFS)	Aircraft Crash

Pilots Kelly and Lynn were killed when their air tanker plane crashed after dropping retardant on a wildfire.

| 8/3/1994 | John Nutter, Sergeant | Career, Age 28 |
| | Louisville Division of Fire, KY | Smoke Inhalation/Burns |

Sergeant Nutter was killed when the roof collapsed under him while performing ventilation at a fire in a storage facility. Sgt. Nutter fell into a storage area where, according to investigators, he was able to force his way into a hallway but was then trapped by interlocking doors and heavy fire conditions. Rescue efforts were hampered by maze-like conditions in the building. Sgt. Nutter was found with a depleted air bottle and dislodged facepiece. He had been exposed to fire conditions that exceeded the protective envelope provided by his turnout gear. Efforts to revive him were unsuccessful. He was wearing an operating PASS device but it was in the "off" position. He died of smoke inhalation and burns.

| 8/7/1994 | Wayne Smith, Captain | Career, Age 37 |
| | New York City Fire Department, NY | Smoke Inhalation/Burns |

Captain Smith was critically injured while conducting search-and-rescue operations on an upper floor of a building when he was trapped by high heat and heavy smoke conditions. Captain Smith was burned over 40 percent of his body and received severe smoke inhalation injuries to his lungs; he died on 10/4/1994. Fourteen other firefighters were injured in the blaze. Initial operations were hampered by a faulty fire hydrant across the street from the building.

8/13/1994	Robert Buc, Flight Crew	Age Unknown
	Joe Johnson, Flight Crew	Age Unknown
	Shawn Zaremba, Flight Crew	Age Unknown
	Hemet Valley Flying Service, CA	Aircraft Crash

Crewmembers Buc, Johnson, and Zaremba were killed when their air tanker crashed en route to a wildfire in Kern County, California.

| 8/13/1994 | James Harvey, Firefighter | Volunteer, Age 39 |
| | Greenwood Fire Department, IN | Heart Attack |

Firefighter Harvey suffered a fatal heart attack during training.

| 8/18/1994 | Herbert Smith, Firefighter | Affiliation and Age Unknown |
| | Shelby Volunteer Fire Department, AL | Heart Attack |

Firefighter Smith suffered a fatal heart attack at a fire.

| 8/18/1994 | Sam McCarty, Firefighter | Affiliation and Age Unknown |
| | Harding County Fire District #2 | Heart Attack |

Firefighter McCarty suffered a heart attack and died while cutting a fire line with a road grater at a grass fire.

| 8/8/1994 | David Castro, Firefighter | Age Unknown |
| | United States Fire Service | Apparatus Rollover |

Firefighter Castro died of traumatic injuries suffered when his water tanker truck overturned on the Quincy–Oroville Highway while en route a wildfire.

| 8/8/1994 | Craig Drury, Sergeant | Volunteer, Age 24 |
| | Highview Fire District, KY | Burns |

Sergeant Drury was caught in a flashover while making entry into a single story house. Sgt. Drury suffered severe burns to his lungs. The fire was started by an arsonist who was disgruntled over an interracial adoption.

| 8/25/1994 | Sydney Bruce Maplesden, Jr., Firefighter | Career, Age Unknown |
| | Oregon Department of Forestry | Overrun by Wildfire |

Firefighter Maplesden died when he was overrun by a wildfire while attempting to cut a fire-break with a bulldozer.

| 8/27/1994 | Paul MacMurray, Firefighter | Volunteer, Age 30 |
| | Hudson Falls Volunteer Fire Department, NY | Trapped |

Firefighter MacMurray responded as part of an engine company to a fire on the first floor of a three-story hotel. Assigned to search for and rescue occupants on the second floor, Firefighter MacMurray and another firefighter successfully evacuated several victims while attempts to extinguish the fire were initiated below them. Upon their return to continue the search, conditions quickly changed from a light haze of smoke to black smoke with high heat conditions. Firefighter MacMurray and his partner became separated in their attempt to locate the stairwell and get out of the building. The other firefighter made several efforts to locate Firefighter MacMurray, but was forced to retreat due to untenable conditions. Several rescue efforts were made, but heavy fire conditions eventually forced the evacuation of all fire personnel to defensive positions as the entire structure burned. His body was recovered the following day. The fire was incendiary in nature.

| 8/27/1994 | Gerald Murray, Assistant Chief | Volunteer, Age 46 |
| | Windham Hose Company #1, NY | MVA |

Assistant Chief Murray died when his personal vehicle crashed while responding to a fire.

| 9/1/1994 | James Harris, Firefighter | Affiliation and Age Unknown |
| | Mechanicville Fire Department, NY | Heart Attack |

Firefighter Harris suffered a fatal heart attack after running to the fire station to respond on a fire call.

| 9/3/1994 | Earl Detty, Sr., Captain | Volunteer, Age 50 |
| | Union Township Fire Department, OH | Heart Attack |

Captain Detty responded to a report of a fire in the woods. Upon arrival, he discovered a legal campfire and then suffered a heart attack and died while walking back to his fire engine.

| 9/6/1994 | Dwight Smith, Firefighter Recruit | Career, Age 34 |
| | Memphis Fire Department, TN | Cardiac |

Firefighter Recruit Smith collapsed in cardiac arrest after a jog during a physical training session. Firefighter Smith was in his fourth week of firefighter recruit class.

| 9/11/1994 | Dewey Henry, Lieutenant | Career, Age 58 |
| | Metro–Dade Fire and Rescue Department, FL | Trauma |

Lieutenant Henry was killed after being trapped under rolls of carpet and debris when the roof collapsed during a fire in a carpet warehouse.

| 9/13/1994 | Gus Fullbright, Chief | Volunteer, Age 49 |
| | Sallisaw Fire Department, OK | Heart Attack |

Chief Fullbright assumed incident command of a fire in a fully involved unattached garage with a vehicle inside. Firefighters had pulled a 1-inch booster line to protect the house exposure and were preparing to deploy larger attack lines in a defensive mode when a loud hiss was heard and an explosion took place. Seven firefighters were burned, including Chief Fullbright who was standing 40–50 feet away wearing only his helmet. Chief Fullbright died of his burn injuries 2 weeks after the incident.

9/15/1994 **Robert L. Johnson, Driver**   **Affiliation and Age Unknown**
          Bureau of Land Management   **Struck by Vehicle**

Mr. Johnson, a driver at the National Interagency Fire Center base camp, was killed when his truck was struck by another vehicle while en route with supplies to a wildfire.

9/22/1994 **James Certain, Captain**   **Volunteer, Age 41**
          Scenic Loop Volunteer Fire Department, TN   **Apparatus Rollover**

Captain Certain was killed when the 1972 Chevrolet 3,000-gallon tanker truck he was driving overturned at an intersection while en route to a house fire. Captain Certain died of his injuries at the scene.

9/23/1994 **John King, Flight Engineer**   **Affiliation and Age Unknown**
          US Army Reserve, ID   **Helicopter Crash**

Flight Engineer King, a civilian serving on a U.S. Army Reserve CH–47D helicopter, died of traumatic injuries when he was struck by a helicopter rotor blade during a crash. The helicopter was attempting to pick up firefighters during the Chicken Complex fire in McCall, Idaho. His unit had been assigned to provide emergency support to the U.S. Forest Service's wildland fire suppression teams.

10/1/1994 **Elias Ovsiovitch**   **Affiliation and Age Unknown**
          Hillcrest Fire Company #1, NY   **Heart Attack**

Mr. Ovsiovitch suffered a fatal heart attack while performing clerical duties at the fire station.

10/6/1994 **Daren Smith, Fire Management Officer**   **Career, Age 23**
          United States Forest Service   **Trauma**

Fire Management Officer Smith of the U.S. Forest Service was killed when he was struck by a falling tree while clearing a fire road.

10/13/1994 **Roy Stephenson, Firefighter**   **Volunteer, Age 42**
          #1 Green Township Volunteer Fire Department, IN   **Heart Attack**

Firefighter Stephenson died after suffering a heart attack while operating a pumper at a working structure fire.

10/29/1994 **Michael DeLane, Firefighter**   **Career, Age 33**
          Newark Fire Department, NJ   **Electrocution**

Firefighter DeLane was climbing down an aerial ladder after roof operations at a two-alarm fire. As DeLane passed a saw to a fellow firefighter, it touched a power line, electrocuting both of them. Firefighter DeLane was killed and the other firefighter injured. One civilian died in the fire, which had been extinguished at the time of the accident.

11/8/1994 **Brian D. Sutton, Sr., Firefighter**   **Career, Age 48**
          Enterprise Fire Company, NJ   **Heart Attack**

Firefighter Sutton was stricken by a heart attack as he was hooking up to a hydrant during pump operations at a house fire.

11/9/1994 **Richard Liddy, Firefighter**   **Volunteer, Age Unknown**
          Basking Ridge Fire Company, NJ   **Heart Attack**

Firefighter Liddy suffered a fatal heart attack while pulling hose at the scene of a house fire.

11/12/1994 **Edward Freeman, Firefighter**   **Career, Age Unknown**
          Memphis Fire Department, TN   **Heart Attack**

Firefighter Freeman suffered a fatal heart attack after returning to his station after an auto fire.

| 11/12/1994 | Jack Lee Hone, Firefighter | Career, Age 31 |
| | Santa Monica Fire Department, CA | Exposure |

Firefighter Hone was exposed to an unknown illness while engaged in emergency support activities. He died 3/18/1996.

| 11/20/1994 | Mary Jo Brown, Firefighter | Career, Age 45 |
| | United States Forest Service | Overrun by Wildfire |

Firefighter Brown died of smoke inhalation after her position was overrun by a rapidly moving wildfire. Two other firefighters deployed their personal shelters and survived the fire.

| 11/23/1994 | Roger Evans, Fire Management Officer | Age Unknown |
| | United States Forest Service | Shot (Accidental) |

Fire Management Officer Evans was killed when a 106mm rifle exploded during training for avalanche control, part of his collateral duties.

| 11/28/1994 | Dale Nelboeck, Firefighter | Affiliation Unknown, Age 38 |
| | Mayfield Heights Fire Department, OH | Heart Attack |

Firefighter Nelboeck collapsed and died of an apparent heart attack as he was carrying a patient to an EMS unit during a rescue call.

| 12/6/1994 | Dwight Burger, Firefighter | Volunteer, Age Unknown |
| | South Danesville Volunteer Fire Department, NY | Apparatus Crash |

Firefighter Burger died in an apparatus crash en route to an emergency.

| 12/10/1994 | Jesse Shockley, Jr., Captain | Career, Age 38 |
| | Fort Bragg Fire Department, NC | Heart Attack |

Captain Shockley died after suffering a heart attack during a training session on ladders.

| 12/24/1994 | Lionel Hoffer, Firefighter | Career, Age 44 |
| | Milwaukee Fire Department, WI | Smoke Inhalation |

Firefighter Hoffer died at a fire in a church when he fell through a hole in the second floor. He had been operating an attack line in the church with an engine company when he went to check a room for fire extension. His crew heard him call for help and found him hanging from a hole in the floor, but they were unable to keep him from falling. He fell approximately 12 feet into the first floor. Rescue crews were alerted by his PASS device and tried to reach him, but they were hampered by the collapsed floor in the front of the building and barred security doors at the rear of the building. Rescue crews eventually fought their way through heavy heat and smoke conditions to Firefighter Hoffer's location, removed him from the building, and administered emergency care. His air supply was exhausted, and his death has been attributed to smoke inhalation.

| 12/26/1994 | Evan Buchholtz, Firefighter | Volunteer, Age 46 |
| | Poy–Sippi Fire Department, WI | Heart Attack |

Firefighter Buchholtz suffered a fatal heart attack while performing duties at a house fire.

| 12/27/1994 | Steven Colona, Chief Engineer | Volunteer, Age 10 |
| | Melfa Volunteer Fire Department, VA | Apparatus Rollover |

Chief Engineer Colona died of traumatic injuries when the tanker truck he was driving overturned on the way to reported fire in a chicken house. Another firefighter was severely injured in the accident. The call turned out to be a false alarm.

| 12/27/1994 | Thomas Wylie, Firefighter | Career, Age 30 |
| | New York City Fire Department, NY | Carbon Monoxide |

Firefighter Wylie suffered carbon monoxide poisoning in a structure fire. He died in 1995.

1/5/1995	Walter Kilgore, Lieutenant	Career, Age Unknown
	Gregory Shoemaker, Lieutenant	Career, Age Unknown
	James Brown, Firefighter	Career, Age Unknown
	Randall Terlicker, Firefighter	Career, Age Unknown
	Seattle Fire Department, WA	Trapped

Four members of the Seattle (WA) Fire Department died when a floor collapsed without warning during a commercial building fire. Lieutenants Kilgore and Shoemaker, and Firefighters Brown and Terlicker died when a modified and unprotected wood floor support failed under heavy fire conditions. Contributing factors to this incident included an unusual and complicated building configuration, companies entering the structure on different levels from different sides of the building, a lack of pre-fire plans, conflicting interpretations of observed fire conditions from different locations on the fireground, personnel not recognizing the significance of their own observations and operations with respect to the overall incident, a lack of progress reports that would have permitted the incident commander to reevaluate his attack plan, and inadequate information passed on to responding companies about an arson threat against the building. The cause of the fire was determined to be arson, and a suspect was charged. As part of the Major Fires Investigation Project, the U.S. Fire Administration conducted a detailed analysis of this incident, and the findings are contained in the report "Four Firefighters Die in Seattle Warehouse Fire."

| 1/7/1995 | Wilbur Pinnell, Firefighter | Affiliation and Age Unknown |
| | Winchester Fire Department, TN | Heart Attack |

Firefighter Pinnell suffered a fatal heart attack while returning from extinguishing a garbage fire that had extended to a commercial occupancy.

| 1/24/1995 | Henry Frizzell, Forestry Technician | Affiliation and Age Unknown |
| | Tennessee Division of Forestry, TN | Heart Attack |

Forestry Technician Frizzell suffered a fatal heart attack while returning from a fire.

| 1/26/1995 | Lathan Grant Smith, Jr., Chief | Volunteer, Age 37 |
| | East Providence Volunteer Fire Department, AL | Heart Attack |

Chief Smith died after suffering a heart attack while fighting a brush fire. Chief Grant was a founder of the East Providence Volunteer Fire Department and a career firefighter with the Talladega Fire Department.

| 1/28/1995 | Victor Melendy, Firefighter | Career, Age 47 |
| | Stoughton Fire Department, MA | Trapped |

Firefighter Melendy died when he was caught in a flashover while searching for victims on the third floor of a rooming house.

1/31/1995	Marcus King, Firefighter	Volunteer, Age 15
	Jared Lee Wright, Firefighter	Volunteer, Age 18
	Claude Volunteer Fire Department, TX	Apparatus Struck by Train

Firefighters King and Wright were killed when a train struck their fire apparatus while they were fighting a brush fire on a railroad right-of-way. Both firefighters suffered severe traumatic injuries and died several days later.

| 2/2/1995 | Ernestine Garcia, Firefighter | Volunteer, Age 56 |
| | Willard Fire Department, NM | Apparatus Rollover |

Firefighter Garcia died when she was thrown from a fire engine in a rollover accident while responding to a brush fire. The vehicle apparently hit a soft shoulder on the side of the road, crossed the road and rolled over. The driver, a 17-year-old junior firefighter, was hospitalized for multiple injuries. He was permitted to drive to fires but prohibited from operating on the fireground until turning 18. The weather was clear, and there was no oncoming traffic at the time of the crash.

2/8/1995	**Glenn Scott, Chief**	**Volunteer, Age Unknown**
	Era Volunteer Fire Department, TX	**Heart Attack**

Chief Scott died from a heart attack while returning from an emergency call.

2/13/1995	**Lisa Batten, Firefighter**	**Affiliation and Age Unknown**
	Gilmer County Fire Department, GA	**MVA**

Firefighter Batten died from traumatic injuries she received in a car accident while returning from a paramedic class.

2/14/1995	**Thomas Brooks, Captain**	**Career, Age 42**
	Patricia Conroy, Firefighter	**Career, Age 43**
	Marc Kolenda, Firefighter	**Career, Age 27**
	Pittsburgh Fire Department, PA	**Asphyxiation**

Three Pittsburgh (PA) Fire Department firefighters died after a stairway collapsed, trapping them in the basement. Captain Brooks, Firefighter Conroy, and Firefighter Kolenda died from asphyxiation when they ran out of air while operating a hose line in the basement. Investigations by the City of Pittsburgh and others after the fire indicated that problems with incident command and accountability were key factors contributing to the firefighters' deaths. Other factors included a possible lack of crew integrity and a failure of the crew to take emergency survival actions that may have helped them escape. All of the deceased firefighters were wearing PASS devices that were found in the "off" position. The fire was incendiary and a suspect was arrested. As part of the Major Fires Investigation Project, the U.S. Fire Administration conducted a detailed analysis of this incident, and its findings are contained in the report, "Three Firefighters Die in Pittsburgh House Fire."

2/14/1995	**Wendell Ayers, Firefighter**	**Volunteer Age 53**
	Pacific Grove Fire Department, CA	**Heart Attack**

Firefighter Ayers collapsed and died of an apparent heart attack while attempting to assist in the rescue of two people on a yacht that had run aground.

2/22/1995	**Shawn O'Brien, Firefighter**	**Volunteer, Age Unknown**
	Franklin Volunteer Fire Department, ME	**Heart Attack**

Firefighter O'Brien collapsed and died of a heart attack after a structure fire.

2/25/1995	**Jimmy Bryant, Chief**	**Affiliation and Age Unknown**
	Indian Field Fire Department, SC	**Heart Attack**

Chief Bryant died of a heart attack after discovering a fire at a campground. The fire was deliberately set.

3/3/1995	**Neil Hyland, Firefighter**	**Volunteer, Age Unknown**
	Massapequa Volunteer Fire Department, NY	**MVA**

Firefighter Hyland died in an automobile accident while responding to a fire call.

3/5/1995	**Raymond Schiebel, Lieutenant**	**Career, Age 49**
	New York City Fire Department, NY	**Heart Attack**

Lieutenant Schiebel went into cardiac arrest while operating at a fire in Brooklyn. He died 2 days later at a hospital. An investigation into the incident revealed that a paramedic allegedly failed to properly intubate Lt. Schiebel during resuscitation efforts, inserting the endotrachial tube into his esophagus instead of into his trachea.

3/8/1995	**Donald Koebel, Firefighter**	**Career, Age 33**
	Johnson County Fire District #2, KS	**Trapped**

Firefighter Koebel died when he became trapped in the basement of a house fire and ran out of air. Firefighter Koebel was part of the initial entry crew attempting to locate the seat of the fire when the floor collapsed beneath him. Heavy smoke and fire conditions prevented other firefighters from rescuing him. The fire originated in the basement of the house.

3/9/1995	**Louis Mambretti, Lieutenant**	**Career, Age 57**
	San Francisco Fire Department, CA	**Burns**

Lieutenant Mambretti died of severe respiratory burns he received after becoming trapped in the garage of a house that was on fire. Lt. Mambretti, the officer on the first arriving engine, had led his crew with the first attack line into the garage when the electrically controlled garage door closed behind them. The fire spread quickly due to 50 mile-per-hour winds, creating heavy fire conditions in the garage and injuring the three firefighters before other crews could breach garage door with axes and saws to pull them out.

3/13/1995	**Bobby Crowe, Forest Ranger**	**Career, Age Unknown**
	Georgia Forestry Commission, GA	**Heart Attack**

Forest Ranger Crowe died of a heart attack after battling a fire in a half acre of wood pallets.

3/15/1995	**Phillip Sherburn, Firefighter**	**Volunteer, Age 57**
	Aumsville Rural Fire Protection District, OR	**Heart Attack**

Firefighter Sherburn died of a heart attack shortly after responding to a house fire. He collapsed while performing water supply operations.

3/18/1995	**Henry Williams, Firefighter**	**Volunteer, Age 53**
	Delran Volunteer Fire Department, NJ	**Heart Attack**

Firefighter Williams died when he suffered a heart attack while taking a firefighter stress test for the New Jersey Forest Service.

3/24/1995	**Donald Kaczka, Fire Engineer**	**Career, Age 57**
	Chicago Fire Department, IL	**Heart Attack**

Fire Engineer Kaczka died after suffering a heart attack at the scene of a rubbish fire at a recycling plant.

3/27/1995	**Dana Morrison**	**Volunteer, Age Unknown**
	Ferry County Fire Protection District, WA	**Heart Attack**

Firefighter Morrison died after suffering a heart attack while operating a hose line at a fire.

3/29/1995	**Norman Prime, Deputy Chief**	**Volunteer, Age Unknown**
	South China Fire Department, MA	**Heart Attack**

Deputy Chief Prime died of a heart attack while fighting a brush fire.

4/2/1995	**James Weaver, Firefighter**	**Volunteer, Age 71**
	Gallupville Volunteer Fire Department, NY	**Heart Attack**

Firefighter Weaver of the Gallupville (NY) Volunteer Fire Department died of a heart attack while performing water supply operations at a brush fire.

4/13/1995	**Herloff "Ted" Hansen III, Firefighter**	**Career, Age 39**
	Hobart Fire Department, IN	**Trapped**

Firefighter Hansen was killed while conducting search operations for two reported trapped victims at a house fire. Firefighter Hanson and another firefighter were operating on the second floor when fire erupted from a concealed space near the stairwell, trapping them on the second floor, where they ran out of air. They were able to find their way to a window where a rescue ladder had been placed. As Firefighter Hansen aided his injured partner through the window and on to the ladder, a flashover occurred and he was killed. Three other firefighters were injured attempting to rescue Firefighter Hansen and his partner.

4/15/1995	**Ronnie Wilson, Forestry Technician**	**Career, Age Unknown**
	Tennessee Division of Forestry, TN	**Heart Attack**

Forestry Technician Wilson suffered a fatal heart attack after returning from the scene of a four-acre brush fire.

4/17/1995	Judith Luster–Stauss, Engineer	Volunteer, Age 50
	Michael Lohbeck, Firefighter	Volunteer, Age 47
	Castella Volunteer Fire Department, CA	Apparatus Rollover

Engineer Luster–Stauss and Firefighter Lohbeck were killed while responding to a barn fire when their tanker truck failed to negotiate a curve and overturned into a creek. Both died of traumatic injuries at the scene of the accident. The firefighters had apparently gone in the wrong direction and were reported to be heading away from the fire at the time of the crash.

| 4/23/1995 | Earl McNeil, Jr., Lieutenant | Volunteer, Age Unknown |
| | Princess Anne Volunteer Fire Department, MD | Heart Attack |

Lieutenant McNeil died of a heart attack after fighting a brush fire. Lt. McNeil was a retired career firefighter with the Boston Fire Department.

| 4/24/1995 | Leroy Cropper, Firefighter | Volunteer, Age 52 |
| | Ocean City Volunteer Fire Department, MD | Heart Attack |

Firefighter Cropper suffered a heart attack after fighting a fire in a hotel. Cropper was hospitalized and died on 4/28/1995.

| 4/30/1995 | Joe Novosad, Firefighter | Volunteer, Age 57 |
| | Porter Volunteer Fire Department Station 2, TX | Heart Attack |

Firefighter Novosad died of a heart attack while responding to a plane crash at Williams Airport.

| 5/5/1995 | Greg Cussen, Firefighter | Volunteer, Age Unknown |
| | Noble Township Volunteer Fire Department, IN | MVA |

Firefighter Cussen was killed when his car collided with another fire department vehicle at the scene of a reported explosion at a school.

| 5/9/1995 | Travis McCormick, Firefighter | Volunteer, Age 68 |
| | New Union Volunteer Fire Department, AL | Heart Attack |

Firefighter McCormick suffered a fatal heart attack at the scene of a mobile home fire.

| 5/12/1995 | Ray Parnell McKay, Jr., Chief | Volunteer, Age 46 |
| | Northeast Lamar County Fire Department, MS | Heart Attack |

Chief McKay suffered a fatal heart attack as he was leaving the fireground.

| 5/12/1995 | Dania Stivers, Firefighter | Volunteer, Age 23 |
| | North Pulaski Fire Protection District, AR | Apparatus Rollover |

Firefighter Stivers was killed when the fire engine in which she was riding overturned, crushing her. The driver of the engine received minor injuries. The engine was on a nonemergency training run and the driver had apparently swerved to avoid an oncoming car. Firefighter Stivers had just completed her 6 month probation period in the department.

| 5/13/1995 | Robert Lapp | Volunteer, Age 44 |
| | Grantsville Volunteer Fire Department, MD | Heart Attack |

Firefighter Lapp died of an apparent heart attack while transporting a patient to the hospital from the scene of a motor vehicle accident.

| 5/24/1995 | Ronald Deer, Firefighter | Volunteer, Age 23 |
| | Wayne Township Fire Department, IN | Apparatus Rollover |

Firefighter Deer was killed when the fire engine in which he was riding overturned. One other firefighter was paralyzed in the incident; two others received minor injuries. Firefighter Deer and the other critically injured firefighter were in the back of the engine and were not wearing seat belts when the accident occurred. The raised roof of the engine separated from the cab when it overturned. The firefighters were thrown from the engine. The engine was responding to a box alarm that turned out to be a false alarm.

6/3/1995	Bradley Hocking, Sr., Chief	Affiliation and Age Unknown
	Pipestone–Berrien Township–Euaclaire Fire Department, MI	Heart Attack

Chief Hocking suffered a heart attack after responding to the scene of a fatal vehicle accident. He was transported by an ambulance to a nearby hospital, where he died on 6/6/1995.

6/5/1995	William Walls, Firefighter	Volunteer, Age 35
	Rock Community Fire Protection District, MO	Heart Attack

Firefighter Walls suffered a fatal heart attack after fighting a fire in a mobile home.

6/6/1995	Peter "Butch" Borwegan, Lieutenant	Career, Age Unknown
	Edison Division of Fire, NJ	Heart Attack

Lieutenant Borwegan was discovered unconscious on the apparatus floor of Edison Fire Station 5. Attempts to resuscitate him were unsuccessful, and he was pronounced dead at a nearby hospital.

6/7/1995	David Barrera, Firefighter	Affiliation and Age Unknown
	Eagle Pass Fire Department, TX	Seizure

Firefighter Barrera died after suffering a seizure while on duty as a dispatcher.

6/10/1995	Richard Hogan, Firefighter	Career, Age Unknown
	Christopher Rizac, Firefighter	Career, Age Unknown
	Sheppard Air Force Base, TX	Burns

Firefighters Hogan and Rizac were killed while fighting a fire at an oil refinery in Addington, Oklahoma. Both had responded in a P–19 crash truck to assist local fire departments with suppression efforts. The fire was located in an oil storage tank and was caused by a lighting strike. Firefighters Rizac and Hogan were killed when several thousand gallons of burning oil boiled over the side of the tank, trapping their crash truck in a mixture of oil and mud. They attempted to flee on foot but were overrun by the flow of oil and died of massive burns.

6/12/1995	Kevin Sutch, Fire Commissioner	Affiliation and Age Unknown
	Levittown Fire Department, NY	Heart Attack

Fire Commissioner Sutch suffered a fatal heart attack while attending the New York State Firemen's Convention in Albany, New York.

6/22/1995	Gary Cockrell, Pilot (Aero Union Corporation)	Career, Age 33
	Lisa Netsch, Pilot (Aero Union Corporation)	Career, Age 31
	Michael Smith, Pilot (United States Forest Service)	Career, Age 48
		Aircraft Crash

Contract Pilots Cockrell and Netsch of Aero Union Corporation, and Pilot Smith of the USDA Forest Service were killed in an aircraft collision over Ramona, CA, when Smith's spotter plane hit the DC–4 air tanker piloted by Cockrell and Netsch. Both planes were on final approach to the airport after returning from dropping fire retardant on a brush fire.

6/22/1995	Carter Martin, Assistant Chief	Volunteer, Age Unknown
	Brookville–Timberlake (VA) Volunteer Fire Department	Drowning

Assistant Chief Martin drowned after he waded into fast moving floodwaters to search three vehicles that had been swept downstream. Chief Martin was swept under water and trapped. He was wearing full protective firefighting turnout clothing while in the water, was not wearing a life vest, and had been tied into a rope that was attached to a fire engine. Two other rescuers were thrown into the water but survived. Martin also served as a career firefighter with the Lynchburg (VA) Fire Department and was an instructor with the Virginia Department of Fire Programs. It was later discovered that the occupants of the vehicles had reached safety on their own prior to the response of emergency personnel.

6/29/1995   **John Woodward, Fire Protection Specialist**   **Career, Age Unknown**
                    New York State Office of Fire Prevention and   **Heart Attack**
                         Control, NY

Fire Protection Specialist Woodward died of a heart attack while conducting a fire inspection.

7/6/1995   **Randy Williford, Lieutenant**   **Career, Age 44**
                    North Little Rock Fire Department, AR   **Heart Attack**

Lieutenant Williford suffered a heart attack after attempting to complete an agility test required for a promotion to the rank of captain. He died of heart failure at a hospital on 7/9/1995.

7/10/1995   **Gary Soupene, Firefighter**   **Volunteer, Age 48**
                    Riley County Rural Fire Department, KS   **MVA**

Firefighter Soupene was killed while responding in his personal vehicle to a reported grass fire. He was slowing down to pick up another volunteer when his vehicle was struck from behind by a car driven by another firefighter responding to the same incident, flipping Firefighter Soupene's car and killing him. The reported fire turned out to be a burning bale of hay.

7/10/1995   **John Schuyler, Firefighter**   **Affiliation and Age Unknown**
                    Weldon Fire Company, PA   **Heart Attack**

Firefighter Schuyler suffered a fatal heart attack while responding on foot to the firehouse after a vehicle fire had been reported.

7/14/1995   **John Weingart, Firefighter**   **Career, Age 56**
                    Detroit Fire Department, MI   **Heart Attack**

Firefighter Weingart suffered a fatal heart attack while hooking up to a fire hydrant at a residential dwelling fire.

7/15/1995   **Edward Pitcher, Assistant Chief**   **Volunteer, Age 39**
                    Sharon County Volunteer Fire Department, CT   **Electrocution**

Assistant Chief Pitcher was electrocuted when he came in contact with a downed power line during the cleanup of debris after a storm.

7/15/1995   **Adam Sorenson**   **Volunteer, Age Unknown**
                    Ruth Volunteer Fire Department, NV   **Apparatus Crash**

Mr. Sorenson was killed when the ambulance he was driving swerved and crashed. The ambulance was not responding to an emergency or transporting any patients when the accident occurred.

7/19/1995   **Arthur Thompson, Fire Commissioner**   **Volunteer, Age Unknown**
                    Freeport Fire Department, NY   **Heart Attack**

Mayor Thompson, Fire Commissioner for the Freeport (NY) Fire Department, suffered a heart attack while en route to a fire. He was transported by an ambulance to a local hospital where he died.

7/20/1995   **Lyle Garlinghouse, Firefighter**   **Career, Age 45**
                    Osceola Fire Department, FL   **Heart Attack**

Firefighter Garlinghouse suffered a fatal heart attack at an EMS incident.

7/21/1995   **Peter Crown, Firefighter**   **Career, Age 39**
                    Honolulu Fire Department, HI   **Helicopter Crash**

Firefighter Crown, a helicopter pilot for the Honolulu (HI) Fire Department was killed when his helicopter crashed in the Koolau Mountains on the island of Oahu. He was conducting a search for a lost hiker and was towing two police officers in the helicopter's basket to the search area when the helicopter crashed in inclement weather. The two police officers were also killed.

7/25/1995   **Mitch Weaver, Firefighter**   **Affiliation and Age Unknown**
                    Tunnelton Volunteer Fire Department, WV   **Heart Attack**

Firefighter Weaver died of a heart attack while at the scene of a vehicle accident.

7/28/1995	Bill Buttram, Firefighter	Volunteer, Age 31
	Josh Oliver, Firefighter	Volunteer, Age 18
	Kuna Rural Fire Department, ID	Overrun by Wildfire

Firefighters Buttram and Oliver were killed when the 1955 brush truck they were driving stalled and they were overrun by a fast moving wildfire.

| 7/30/1995 | William Luker, Chief | Volunteer, Age 29 |
| | Cedar Creek Volunteer Fire Department, MS | Apparatus Rollover |

Chief Luker was killed when the 1972 3,500-gallon tanker truck he was driving overturned en route to a barn fire, ejecting him from the truck. He was not wearing a seat belt at the time of the accident.

| 8/1/1995 | June Fitzpatrick | Volunteer, Age Unknown |
| | Rocky Point Fire District, NY | Stroke/CVA |

Ms. Fitzpatrick died of a stroke while en route to the fire station for an emergency.

| 8/5/1995 | William Marks, Firefighter | Volunteer, Age Unknown |
| | Munhall Fire and EMS, PA | Apparatus Rollover |

Firefighter Marks was killed when the engine on which he was riding jumped a curb and overturned en route to an electrical fire in a row house. He was riding on the tailboard of the engine at the time of the accident. Several other firefighters were seriously injured.

| 8/7/1995 | Eric Mangieri, Firefighter | Affiliation and Age Unknown |
| | New Kensington Fire Company, PA | Trapped |

Firefighter Mangieri was killed when he became trapped while trying to escape from a house fire during a flashover.

| 8/16/1995 | Christopher Garneau, Firefighter | Volunteer, Age 17 |
| | Warrenton Volunteer Fire Company, VA | Heart Attack |

Firefighter Garneau died of a heart attack while responding to an emergency incident. An autopsy revealed that Firefighter Garneau, age 17, had an enlarged heart.

| 8/20/1995 | Raymond Trygar, Battalion Chief | Career, Age Unknown |
| | California Department of Forestry, CA | Exposure |

Battalion Chief Trygar was exposed to an unknown illness on 8/20; he died 12/20/1995.

| 8/21/1995 | Bruce Cormican, Firefighter/EMT | Affiliation and Age Unknown |
| | Black River Falls Fire Department, WI | Drowning |

Firefighter–EMT Cormican drowned while conducting a body recovery in a creek for a drowning victim. Firefighter Cormican and two other members of the Black River Fall dive team were trapped in a hydraulic created by a small waterfall while wading into the creek and searching for the victim's body. Shore personnel were able to remove the three rescuers with a pike pole, but they could not revive Firefighter Cormican.

| 8/24/1995 | Corey Berggren, Rescue Diver | Affiliation and Age Unknown |
| | Knoxville Volunteer Rescue Squad, TN | Asphyxiation (while diving) |

Rescue Diver Berggren died while conducting dive operations in a quarry to recover the body of a drowning victim. He had apparently inhaled the wrong mixture of gas from his scuba tank during the 200-foot dive, asphyxiating him.

| 8/27/1995 | James Greg Hinson, Firefighter | Volunteer, Age Unknown |
| | Mebane Fire Department, NC | Drowning |

Firefighter Hinson drowned after rescuing a man from a vehicle on a flooded highway. Firefighter Hinson and two other firefighters had tied themselves into a rope that was attached to a haul line on a fire department vehicle. As he reached the car and pulled the occupant out, the car was swept into the flooded creek channel. Firefighter Hinson and the other firefighters slipped into the current, and he became trapped on a guide wire to a telephone pole when other rescuers tried to haul them out of the water. The two other firefighters were rescued. A rescue diver from another department entered the water to reach the car driver, and they were both swept into a tree, where they were later rescued by boat. Firefighter Hinson was eventually pulled from the water, but attempts to resuscitate him were unsuccessful.

| 8/31/1995 | **Martin Kautz, Firefighter** | **Volunteer, Age 35** |
| | **Brush Volunteer Fire Department, CO** | **MVA** |

Firefighter Kautz was killed in a vehicle accident while en route to a medical emergency. Firefighter Kautz and three other firefighters responded in his minivan from the fire station to a report of a child choking. They were struck en route by a car driven by an emergency medical technician who had run a stop sign while en route to the ambulance quarters to respond to the same call. Firefighter Kautz was killed and the three firefighters were injured.

| 9/6/1995 | **George Peters, Fire/Police Officer** | **Volunteer, Age Unknown** |
| | **Eflinwild Volunteer Fire Company, PA** | **Heart Attack** |

Fire/Police Officer Peters suffered a fatal heart attack at the scene of a multialarm fire.

| 9/8/1995 | **John Pache, Past Chief** | **Volunteer, Age Unknown** |
| | **Aviation Volunteer Fire Company No. 3, NY** | **Heart Attack** |

Past Chief Pache suffered a fatal heart attack while en route to a structure fire.

| 9/13/1995 | **John McCroden, Captain** | **Volunteer, Age 47** |
| | **City of Geneva Fire Department, OH** | **Heart Attack** |

Captain McCroden died when he suffered a heart attack while ventilating the roof at a fire in a single-family house.

| 9/16/1995 | **Eric Schaefer, Firefighter** | **Career, Age 25** |
| | **Baltimore City Fire Department, MD** | **Trauma** |

Firefighter Schaefer was killed when a 1 1/2 foot thick granite wall collapsed on him while he was engaged in forcible entry at a multiple-alarm fire in an old warehouse that had been converted to a business occupancy.

| 9/17/1995 | **Gene Schubert, Firefighter** | **Affiliation and Age Unknown** |
| | **Harriman Fire Department, TN** | **Heart Attack** |

Firefighter Schubert suffered a fatal heart attack at the scene of a fire.

| 9/19/1995 | **Ray Lencioni, Captain** | **Affiliation and Age Unknown** |
| | **Colma Fire District, CA** | **Heart Attack** |

Captain Lecnioni suffered a fatal heart attack while at a medical call.

| 9/23/1995 | **Frederick Fairweather, First Assistant Chief** | **Volunteer, Age 48** |
| | **Bullville Fire Company, NY** | **Heart Attack** |

First Assistant Chief Fairweather suffered a fatal heart attack while marching with the department in the Bullville parade. He was a retired firefighter with the Newburgh Fire Department.

| 9/26/1995 | **Thomas O'Boyle, Lieutenant** | **Career, Age 55** |
| | **Chicago Fire Department, IL** | **Heart Attack** |

Lieutenant O'Boyle suffered a fatal heart attack at a fire in a furniture warehouse. Lt. O'Boyle, aide to the Fire Commissioner, had just exited the structure after checking on the progress of firefighters inside when he collapsed.

| 9/28/1995 | **John Fisher, Sergeant** | **Affiliation and Age Unknown** |
| | **Greensburg Volunteer Fire Department, PA** | **Heart Attack** |

Sergeant Fisher suffered a fatal heart attack while responding on foot to an emergency call for a pedestrian struck by a train.

9/30/1995   **Richard Washburn, Instructor**   **Affiliation and Age Unknown**
                          Kentucky Tech Fire/Rescue Training, KY   **Unknown**

Instructor Washburn collapsed and died while he was teaching a confined space entry and rescue course at a regional fire school.

10/8/1995   **Peter McLaughlin, Firefighter**   **Career, Age 31**
                          New York City Fire Department, NY   **Trapped**

Firefighter McLaughlin died from burn injuries and smoke inhalation while performing a search on the fourth floor at a fire in a tenement building. The fire broke through the ceiling and engulfed the apartment. A window gate blocked Firefighter McLaughlin's route of escape. The fire started on a mattress in a fourth floor bedroom. The building had been previously cited for over 170 fire code violations.

10/25/1995   **John Riggins, Jr., Corporal**   **Career, Age 53**
                          Indianapolis Fire Department, IN   **Heart Attack**

Corporal Riggins suffered a fatal heart attack after performing roof ventilation at a house fire.

10/28/1995   **James "Frank" Ainsworth, Past Chief**   **Affiliation and Age Unknown**
                          Friendship Fire Company, WV   **Heart Attack**

Past Chief Ainsworth suffered a fatal heart attack at a fundraising event for the fire company. He was treasurer and served on the Board of Directors of the department at the time of his death.

10/29/995   **Stephen Sulzinski, Firefighter**   **Volunteer, Age 66**
                          Hicksville Fire Department, NY   **Heart Attack**

Firefighter Sulzinski suffered a fatal heart attack while attending a fire service activity in Albany.

11/7/1995   **Walter Augustin, Assistant Chief**   **Volunteer, Age Unknown**
                          Congers Fire Department, NY   **Heart Attack**

Assistant Chief Augustin suffered a fatal heart attack while marching in the Congers parade with the department.

11/9/1995   **John Haviar, Firefighter**   **Career, Age Unknown**
                          Industrial Fire Brigade   **Asphyxiation**

Firefighter Haviar, an industrial fire brigade member at a Reynolds Aluminum plant, was killed when he entered an oxygen deficient atmosphere in an excavation pit without breathing apparatus to attempt the rescue of three workers. Argon gas was accidentally pumped into the pit, displacing the oxygen and trapping the workers. Firefighter Haviar was overcome and was killed along with two of the workers.

11/11/1995   **Thomas Buff, Jr., Chief**   **Affiliation and Age Unknown**
                          Blaney Volunteer Fire Department, SC   **Struck by Vehicle**

Chief Buff was killed when a passing police cruiser struck him. The fire department had just finished extinguishing a vehicle fire on an interstate highway when a minor accident occurred in the opposing lanes of traffic. Chief Buff started across the road to check on the occupants of the cars when he was hit, receiving multiple traumatic injuries.

11/19/1995   **David Harness, Assistant Chief**   **Volunteer, Age 45**
                          Hanna Township Volunteer Fire Department, IN   **Struck by Vehicle**

Assistant Chief Harness was killed when a vehicle at the scene of an emergency struck him.

11/25/1995   **Michael Canonico, Firefighter**   **Volunteer, Age 52**
                          Andover Township Fire Department, NJ   **MVA**

Firefighter Canonico was killed while responding on his motorcycle to a report of a furnace fire. He died when he attempted to pass a pickup truck on the right and the truck made a right turn, striking him.

12/10/1995	Henry W. Howe, Firefighter	Volunteer, Age 50
	Brownsville Rural Fire District, OR	Trauma

Firefighter Howe was struck by a vehicle while clearing a downed tree from a road.

12/14/1995	James Shue, Captain	Volunteer, Age Unknown
	Locke Township Volunteer Fire Department, NC	Apparatus Rollover

Captain Shue was killed when the engine he was driving overturned en route to an odor of smoke call, which turned out to be a false alarm. Two other firefighters were injured.

12/31/1995	John Clancy, Lieutenant	Career, Age 35
	New York City Fire Department, NY	Burns

Lieutenant Clancy was killed when the floor collapsed beneath him at a fire in an abandoned residential building as he entered to conduct a search for occupants. Lt. Clancy fell into the basement where he died of burns.

1/5/1996	James B. Williams, Firefighter	Career, Age 38
	New York City Fire Department, NY	Burns

Firefighter Williams died from burns sustained during a two-alarm fire at an apartment building in Queens, New York. Unaware that the occupants of the apartment had already left, he and four other firefighters were searching for victims and fighting the fire when they were engulfed in flames after breaking through a door.

1/5/1996	William R. Favinger, Sr., Firefighter	Volunteer, Age 55
	West End Fire Company, PA	Heart Attack

Firefighter Favinger suffered a fatal heart attack while returning from an automatic alarm. He collapsed in the station while filling out the roster. CPR was immediately started and paramedics were called.

1/6/1996	Guy R. Pollard, Firefighter	Volunteer, Age 64
	Owego Fire Department, NY	Heart Attack

Firefighter Pollard suffered a fatal heart attack while performing pump operations at a suspected house fire on a mutual call with the Owego Fire Department. After determining that the house's chimney was stuffed, trucks began preparing to leave when Firefighter Pollard suffered the heart attack. He was transported to Wilson Memorial Regional Medical Center where he was pronounced dead on arrival.

1/7/1996	Thomas Dorr, Firefighter	Volunteer, Age 50
	Pleasantville Fire Department. NY	Stabbed

Firefighter Dorr died from multiple stab wounds while responding to the station on foot during a snowstorm. He was walking to the firehouse for storm duty and was attacked en route.

1/7/1996	Willard Hopler, Firefighter	Volunteer, Age 59
	Rockaway Borough Fire Department, NJ	Heart Attack

Firefighter Hopler suffered a massive heart attack while operating an aerial apparatus at the scene of a chimney fire. Despite resuscitative efforts by his crew, he was pronounced dead on arrival at the hospital

1/13/1996	Dale Zimmerman, Chief	Volunteer, Age 40
	Pecatonica Fire Protection District, IL	Asphyxiation

Chief Zimmerman died in an attempt to rescue two men who were overcome by fumes in a grain bin. During the rescue, alarm bells sounded warning that another firefighter's SCBA was low. Chief Zimmerman went over to change the tank, but his mask fogged up. He took off his mask in order to change the tank. The other firefighters then left, but saw Chief Zimmereman having problems in his attempt to rescue the two men. The two men were rescued, but Chief Zimmerman eventually died from asphyxia from the carbon monoxide poisoning.

| 1/18/1996 | Marcel Glenn, Firefighter | Career, Age 34 |
| | Cairo Fire Department, GA | Cardiac |

Firefighter Glenn died while fighting a structure fire. He was ventilating a house fire by breaking the windows with a fire hose. After he had ventilated two windows, he turned around and collapsed. EMS was called and he was taken to the hospital where he was pronounced dead due to cardiac arrest.

| 1/19/1996 | Reed Morton, Sr., Fire Police | Volunteer, Age 78 |
| | Citizens Hose Company #1, PA | Heart Attack |

Firefighter Morton suffered a heart attack while directing traffic and assisting evacuees at a fire.

| 1/19/1996 | Robert Haggadone | Volunteer, Age 43 |
| | Wildwood Fire Association, MI | Struck by Vehicle |

Firefighter Haggadone was struck by a passing motorist while operating a pumper at a house fire. He died after being in a coma for 7-1/2 months.

| 1/22/1996 | Jerald Dibbles, Firefighter Recruit | Career, Age 23 |
| | Dallas Fire Department, TX | Sickle Cell |

Firefighter Recruit Dibbles died during his second day at the training academy. He had a pre-existing condition of sickle cell anemia (trait). Because of this condition, he went into a state of rhabomyalisis (internal heating and buildup of acid in the heart muscle). This condition caused several major organs to fail resulting in death.

| 1/26/1996 | Dale Burkhalter, Firefighter | Volunteer, Age 27 |
| | Livingston Fire Department, TX | MVA |

Firefighter Burkhalter died in a car accident while returning from a fire incident. The accident occurred at 4 a.m. at a dark and unlighted intersection. The road conditions was wet, and there were patches of fog. In an attempt to cross a major highway, the Firefighter Burkhalter's car was struck on the driver's side.

| 1/31/1996 | Marvin Mathis, Firefighter | Volunteer, Age Unknown |
| | Lake Murray Village Fire Department, OK | Heart Attack |

Firefighter Mathis drove a tanker to the fire scene and suffered a fatal heart attack on arrival.

| 2/1/1996 | Steven Gushiken, Firefighter | Career, Age 52 |
| | Kauai County Fire Department, HI | Unknown |

Firefighter Gushiken got up in the morning (4:30 a.m.) while still on duty and went for a walk in the park adjacent to the station as part of his normal morning routine. When the rest of the shift woke up, they found him unconscious on the ground. Attempts to revive him were unsuccessful.

| 2/5/1996 | Louis Valentino, Firefighter | Career, Age 37 |
| | New York City Fire Department, NY | Trapped |

Firefighter Valentino became trapped when the roof of an Auto Body Shop in East Flatbush–Brooklyn collapsed. Fifteen other firefighters were injured at this three-alarm blaze. Firefighter Valentino died less than an hour after the fire started at about 3:40 p.m.

| 2/5/1996 | Corey Morgan, Firefighter | Volunteer, Age 16 |
| | Clarksville Fire Department, VA | MVA |

Firefighter Morgan died in a motor vehicle accident while responding to a fire call.

| 2/11/1996 | Michael McLaughlin, Firefighter | Volunteer, Age 59 |
| | Ridgefield Boro Fire Department, NJ | Trauma |

Firefighter McLaughlin died of an apparent heart attack after arriving on the scene of a small fire in a laundromat. He experienced head trauma when he fell on the scene and knocked his head against the fire engine. This trauma resulted in cardiac arrest.

2/11/1996   **Raymond Vinson, Firefighter**   **Volunteer, Age 61**
            Enville Fire Department, OK   **Heart Attack**

Firefighter Vinson fought a grass fire for approximately 7 hours in the morning, when he was called out again for another grass fire. This incident lasted about 5 hours. He died of a heart attack after returning from the incident.

2/23/1996   **Nathaniel Quinn, Firefighter**   **Volunteer, Age 66**
            I.X.L. Fire Department, OK   **Cardiac**

Firefighter Quinn went into cardiac arrest while fighting a wildland fire near Okemah. The wild-fires blackened up to 30,000 acres and destroyed 43 homes in 10 Oklahoma counties.

2/24/1996   **Francis Ploeger, Firefighter**   **Volunteer, Age 45**
            Ash Township Volunteer Fire Department, MI   **Heart Attack**

Firefighter Ploeger arrived at the scene of a two-alarm barn fire. While pulling a hose from the fire truck, he collapsed from a heart attack. CPR was initiated at the scene, and he was taken to a hospital where he was pronounced dead later that evening.

3/2/1996   **Leonardo Maguidad, Firefighter**   **Volunteer, Age 29**
            Allentown Road Volunteer Fire, MD   **Cardiac**

Firefighter Maguidad suffered a cardiac arrest at the station while on duty.

3/7/1996   **Dennis McGary, Firefighter**   **Volunteer, Age 46**
            Tomah Volunteer Fire Department, WI   **Heart Attack**

Firefighter McGary suffered a fatal heart attack after returning from a house fire. After returning, he was putting away equipment and preparing firehouse items when the heart attack occurred.

3/8/1996   **Vinton Durflinger, Firefighter**   **Volunteer, Age 72**
            Alexandria Volunteer Fire Department, NE   **Heart Attack**

Firefighter Durflinger collapsed and died due to a heart attack after checking out a suspected house fire.

3/13/1996   **Norman Manka, Firefighter**   **Volunteer, Age 44**
            Golden City Volunteer Fire Department, MI   **Heart Attack**

Firefighter Manka was operating the pump at a grass fire when he collapsed and died due to a heart attack.

3/18/1996   **Frank Young, Firefighter**   **Career, Age 38**
            John Hudgins, Jr., Firefighter   **Career, Age 32**
            Chesapeake Fire Department, VA   **Trapped**

Firefighters Young and Hudgins died while battling a blaze in the Advanced Auto Parts store. Both firefighters became trapped by fire when the truss roof collapsed on top of them. The fire-fighters were found in the rear of the structure some time later.

4/7/1996   **Robert Duvall, Firefighter**   **Volunteer, Age 39**
            Granville Fire Protection District, IL   **Heart Attack**

Firefighter Duvall suffered a fatal heart attack while fighting a house fire in Hennepin, IL.

4/8/1996   **Norman Adams, Firefighter**   **Volunteer, Age 38**
            Almena V. Fire Department, KS   **Asthma Attack**

Firefighter Adams died from an asthma attack after engaging in support duties for 9-1/2 hours at an industrial fire in a plant that makes aluminum products.

4/8/1996   **Jonathan C. Boster, Firefighter**   **Volunteer, Age 19**
            Grant County Fire District 5, WA   **Apparatus Rollover**

Firefighter Boster and another firefighter were responding to a reported mobile home fire when they rounded a corner too quickly and the tanker they were in rolled onto its side. Firefighter

Boster was killed. The other firefighter was treated for broken ribs and other minor injuries. It was not reported whether the firefighters were wearing seat belts.

| 4/10/1996 | **Terry Leasher, Firefighter** | **Volunteer, Age 41** |
| | **Harlan Township Fire & Rescue, OH** | **MVA** |

Firefighter Leasher died of internal injuries due to a motor vehicle accident. He was on his way to the station to perform truck inspections.

| 4/19/1996 | **Donald Collins, Firefighter** | **Career, Age 62** |
| | **Schenectady City Fire Department, NY** | **Cardiac** |

Firefighter Collins was hooking up a hose at the scene of a vacant house fire around 2 a.m. when he went into cardiac arrest. He was taken to the hospital and pronounced dead.

| 4/19/1996 | **Mathew Hatcher, Firefighter** | **Volunteer, Age 32** |
| | **Wayne Fire Department, OK** | **Struck by Apparatus** |

Firefighter Hatcher died of abdominal injuries after being pinned between two fire trucks at a grass fire. He was at the rear of one truck starting the pump when a second truck struck him.

| 4/23/1996 | **John Goessling, Firefighter** | **Career, Age 43** |
| | **Omaha Fire Department, NE** | **Trauma** |

Firefighter Goessling was killed when the roof collapsed on him at a four-alarm fire in a commercial building, the Dollar General store. A 15-year-old was arrested and charged with arson.

| 4/23/1996 | **Mark Clark, Firefighter** | **Career, Age 45** |
| | **Division of Forestry & Wildlife, HI** | **Heart Attack** |

Firefighter Clark died while participating in a chainsaw (tree felling) training class. He was clearing brush around a tree when he put his saw down, collapsed, and died of a heart attack.

4/24/1996	**Stanley Adams, Captain**	**Career, Age 45**
	Don Moree, Captain	**Career, Age 49**
	Willie Craft, District Chief	**Career, Age 48**
	Rick Robbins, District Chief	**Career, Age 47**
	Jackson Fire Department, MS	**Shot**

Captain Adams, Captain Moree, District Chief Craft, and District Chief Robbins were shot to death during a meeting of district chiefs. They were killed by a disgruntled firefighter who went on a rampage killing five people including his wife. Two other people were also injured in this incident.

| 4/26/1996 | **Robert Pemberton, Firefighter** | **Volunteer, Age 55** |
| | **Antioch Volunteer Fire Department, AR** | **Apparatus Rollover** |

Firefighter Pemberton was killed in an apparatus accident while en route to a reported structure fire. He was ejected from the driver's seat after the truck failed to negotiate a turn and then overturned several times. He was pronounced dead at the scene.

| 5/1/1996 | **Robert Hamler** | **Career, Age 34** |
| | **Atlanta Fire Department, GA** | **Stroke** |

Firefighter Hamler suffered a stroke at the fire station. He was inspecting fire hydrants when he started to feel poorly. He was then taken to the hospital where he died.

| 5/26/1996 | **Kevin Malone, Assistant Chief** | **Volunteer, Age 31** |
| | **Mahwah Township Fire Department, NJ** | **Heart Attack** |

Assistant Chief Malone died of a heart attack at home after returning from a false alarm. He had complained of not feeling well earlier in the day, and 2 days earlier he had fought a separate fire where he inhaled a large amount of smoke.

| 5/30/1996 | **William Frank, Firefighter** | **Volunteer, Age 73** |
| | **Camp Hill Fire Department, PA** | **Heart Attack** |

Firefighter Frank had a heart attack after returning from a heat exchanger fire at a mall.

6/9/1996   Michelle Smith, Firefighter   Career, Age 23
                  Globe Ranger District (USFS), AZ   Heat Exhaustion/Dehydration

Firefighter Smith disappeared during a training run and was found dead 26 hours later. According to the autopsy report, she died of heat exhaustion and dehydration. There was no sign of struggle or foul play.

6/19/1996   Rex Hoad, Firefighter   Volunteer, Age 43
                  Cameron Fire Department, NY   MVA

Firefighter Hoad died from injuries from a motor vehicle accident that occurred while returning from a service call.

6/23/1996   Lee Allen Steele, Firefighter   Volunteer, Age 22
                  Monte Jason Harmon, Firefighter   Volunteer, Age 19
                  Poplar Springs Fire Department, SC   Apparatus Rollover

Firefighters Steele and Harmon were killed in an apparatus accident while responding to a call that turned out to be a false alarm. The driver lost control of the truck, and it ran off the road, overturned, and hit a tree.

6/24/1996   Ronald Lupo, Firefighter   Volunteer, Age 32
                  Dillon County Fire Department, Dillon, SC   MVA

Firefighter Lupo was responding to a field fire at approximately 7:10 p.m. While en route, his vehicle was struck on the right front side by an oncoming van. He died from internal injuries later that night at a hospital.

6/29/1996   Robert Bibbee, Firefighter   Volunteer, Age Unknown
                  Elizabeth Volunteer Fire Department, WV   Heart Attack

Firefighter Bibbee was hauling drinking water to families and homes in a rural area to raise funds for a fire department event when he suffered a heart attack.

7/4/1996   William L. Parsons, Captain   Volunteer, Age 32
                  Cameron Volunteer Fire Department, WV   Trauma

Captain Parsons, a licensed pyrotechnician, was killed at the annual 4th of July fireworks display that is sponsored by the Cameron Fire Department. A 6-inch round prematurely detonated on the ground causing a piece of metal/wood to strike Captain Parsons in the head. His brother was also injured.

7/11/1996   Bruce Lindner, Firefighter   Volunteer, Age 48
                  Relief Hose Company #2, NJ   Cardiac

Firefighter Lindner died from cardiac arrest during a vehicle extrication drill.

7/11/1996   George Crane, Jr., Firefighter   Volunteer, Age 23
                  Morgan County Fire Department, GA   MVA

Firefighter Crane died in a motor vehicle accident while responding in his personal vehicle to an emergency call.

7/13/1996   Arthur Petit, Firefighter   Career, Age 59
                  Holyoke Fire Department, MA   Cardiac

Firefighter Petit died due to cardiac arrest while searching for victims at a multifamily dwelling fire. The firefighter's crew was ordered to search the interior of the third floor (fire floor). While searching and ventilating, Firefighter Petit collapsed on the porch of one of the apartments. He was not revived despite the attempts of fellow firefighters who performed CPR immediately.

7/21/1996   Donald Raibley, Firefighter   Volunteer, Age 19
                  Pigeon Township Volunteer Fire Department, IN   Drowning

Firefighter Raibley was responding to a residential house fire at 4 a.m. when he had a seizure and drove his car over a dam into a lake, where he drowned.

| 7/27/1996 | Kris Sherman, Firefighter | Volunteer, Age 36 |
| | Sligo Volunteer Fire Department, PA | Apparatus Rollover |

Firefighter Sherman died from injuries resulting from an overturned pumper during a response to an incident.

| 7/28/1996 | George Guyer, Firefighter | Volunteer, Age 38 |
| | Stokes–Rockingham V.F.D, NC | MVA |

Firefighter Guyer was responding in his personal vehicle to a transformer fire when his truck hydroplaned on the wet road and collided with an oncoming truck.

| 8/6/1996 | John William Swan II, Firefighter | Volunteer, Age 18 |
| | Township Volunteer Fire Department, IN | Shot |

Firefighter Swan responded to a motor vehicle accident involving a car and a motorcycle. The driver of the motorcycle ran into the boat that the car was towing. The driver of the car proceeded to shoot the driver of the motorcycle, two bystanders, and the firefighter who arrived on the scene.

| 8/8/1996 | Norman Ray, Firefighter | Volunteer, Age Unknown |
| | Fruitland Fire Department, UT | Cardiac |

Firefighter Ray suffered cardiac arrest due to overexertion at a grass fire.

| 8/8/1996 | Floyd Birchmore, Firefighter | Volunteer, Age 53 |
| | Addison Fire Department, VT | Heart Attack |

A fire broke out in a barn owned by Firefighter Birchmore. He called the fire department and started to lead the animals out of the barn. The first engine arrived and Firefighter Birchmore began pulling the hose off the truck. Shortly after, he collapsed and died of a heart attack.

| 8/8/1996 | William McGinnis, Firefighter | Volunteer, Age 38 |
| | Surfside Volunteer Fire Department, TX | Electrocution |

Firefighter McGinnis was the lone responder to an electrical pole fire that occurred midday. As he began deploying a hoseline, the pole broke in half, pulling the charged power lines down and electrocuting him.

| 8/12/1996 | Walter L. Bricker, Deputy Chief | Volunteer, Age 65 |
| | Metal Township Volunteer Fire & Ambulance Company, PA | Heart Attack |

Chief Bricker suffered a fatal heart attack while assisting an ambulance crew with patient care. He was pronounced dead upon arrival at the hospital.

| 8/21/1996 | Leonard Coulter, Firefighter | Volunteer, Age 67 |
| | Gerton Fire Department, NC | Cardiac |

Firefighter Coulter suffered cardiac arrest while responding to a motor vehicle accident.

| 8/24/1996 | Leslie Hendricks, Deputy Chief | Career, Age 59 |
| | Union Township Fire Department, NJ | Heart Attack |

Deputy Chief Hendricks died as a result of cardiac arrest that was connected to an exposure at an earlier fire incident at a Burger King. He was supervising a crew when a cloud of gas vapors engulfed him. He began having trouble breathing and was sent to the hospital. Deputy Chief Hendricks was discharged after 2 days, returned home and died 10 days later.

| 8/25/1996 | John Gray, Firefighter | Career, Age 50 |
| | Bureau of Land Management, NV | Heart Attack |

Firefighter Gray was repairing a water tender between fires when he died of a heart attack. After he collapsed, a fellow crewmember began CPR and called an ambulance.

| 8/26/1996 | Lawrence Roche, Lieutenant | Career, Age 46 |
| | Harahan Fire Department, LA | Heart Attack |

Lieutenant Roche had a heart attack at the scene of a structure fire.

8/27/1996   **Robert Wallingford, Captain**   **Volunteer, Age 47**
                    South Portland Fire Department, ME   **Heart Attack**

Captain Wallingford died from a heart attack while directing engine company operations at the scene of a four-alarm fire in a welding supply company.

9/4/1996   **Richard Dorsey, Firefighter**   **Volunteer, Age 19**
                    Bahama Volunteer Fire Department, NC   **MVA**

Firefighters were responding to a call when a tree fell across the roadway and struck their brush truck. The accident killed Firefighter Dorsey and injured one other.

9/10/1996   **Bruce Honstain, Assistant Chief**   **Volunteer, Age 43**
                    Powell Volunteer Fire Department, WY   **Electrocution**

Assistant Chief Honstain was attempting to rescue his son from a motor vehicle accident when they were both electrocuted and died.

9/14/1996   **Sam Strall, Firefighter**   **Volunteer, Age 35**
                    Rising Sun Fire Department, MD   **Heart Attack**

Firefighter Strall collapsed and died of a heart attack during a fundraiser at the firehouse.

9/18/1996   **Jeffrey Renner, Firefighter**   **Volunteer, Age 35**
                    Baltic Fire & Rescue Department, OH   **Heart Attack**

Firefighter Renner had arrived at his regular job, when he was informed of a fire in the paint shed. He was leaving to drive to the station to get his gear when he suffered a heart attack as he was getting to his car.

9/18/1996   **Henry Scott, Firefighter/Paramedic**   **Career, Age 36**
                    Springdale Fire Department, OH   **Heart Attack**

Firefighter/Paramedic Scott suffered a fatal heart attack while at a live burn training exercise.

9/20/1996   **William Reid, Firefighter**   **Career, Age 42**
                    Birmingham Fire and Rescue Service, AL   **Cardiac**

Firefighter Reid died of cardiac arrest at required annual fitness test (running and walking).

9/21/1996   **Martin Doherty, Lieutenant**   **Career, Age 63**
                    Chicago Fire Department, IL   **Heart Attack**

Lieutenant Doherty, a 35-year veteran of the Chicago Fire Department, suffered a fatal heart attack while on duty at the fire station.

9/29/1996   **Kevin Reveal, Firefighter**   **Volunteer, Age 31**
                    Herrin Fire Department, IL   **Trauma**

Firefighter Reveal died while fighting a fire at a commercial two-story, vacant, boarded-up, wood-frame structure. He was opening up a boarded window that was opposite from where the fire was located when the wall collapsed, killing him and injuring several others.

10/12/1996   **Clark Derryberry, Firefighter**   **Volunteer, Age 19**
                    Mt. Pleasant Rural Fire, TN   **MVA**

Firefighter Derryberry died in a motor vehicle accident while returning home from a barn fire. The barn fire was the last one out of a series of four.

10/13/1996   **Keith Boudoin, Firefighter**   **Career, Age 41**
                    Jefferson Parish EBC Fire Dept., LA   **Heart Attack**

Firefighter Boudoin was preparing to enter a structure fire for the third time to look for trapped victims when he suffered a fatal heart attack. He was immediately taken to the hospital where he was pronounced dead.

10/15/1996   **Karl Schmidt, Assistant Chief**   **Volunteer, Age 40**
                    Cowlesville Volunteer Fire Company, NY   **Cardiac**

Assistant Chief Schmidt died from an apparent heart attack after attending an EMS training event. The chief returned home from the event, and then decided to drive himself to the hospital but died on the way.

10/18/1996 Martha Ann Bice, Firefighter Volunteer, Age 59
West Etowah VFD, AL Surgical Complications

Firefighter Bice was cutting firebreaks at a brush fire when she experienced chest pain and collapsed. She was taken to the hospital and doctors determined that she had suffered a heart attack. She underwent triple bypass surgery, returned home and died several days later from complications.

10/19/1996 Eugene Bauerlien, Firefighter Volunteer, Age 72
Westminster Fire Department, MD Heart Attack

Firefighter Bauerlien suffered a fatal heart attack at the fire station. He had been on duty all morning at the station cooking for the fire department fundraiser. At 11:00 a.m., he left to direct traffic for a college homecoming. This was part of his duty for the fire/police, a division of the fire department. He then returned to the fire department for breakfast and went into cardiac arrest.

10/24/1996 Jackson Capps, Captain Volunteer, Age 25
Glassy Mountain Fire Department, SC MVA

Captain Capps died in a motor vehicle accident when his fire truck was struck by a dump truck while responding to a grass fire. He had just finished working the A shift at the electric plant when the call for a grass fire came out. After picking up a fire truck at the fire department, he pulled into a dump truck's path after driving less than 100 yards. The call turned out to be a false alarm.

10/24/1996 Jack Grosse, Firefighter Career, Age 53
Cedar Falls Fire Department, IA Cardiac

Firefighter Grosse suffered cardiac arrhythmia and died while asleep in his quarters.

10/26/1996 Albert DeFlumere, Firefighter Volunteer, Age 49
Blauvelt Volunteer Fire Company, NY Smoke Inhalation

Firefighter DeFlumere died of smoke inhalation at a residential structure fire after he returned inside to rescue his son.

10/26/1996 Frank Gilbert, Jr., Firefighter/Paramedic Career, Age 39
Portage Fire Department, IN MVA

Firefighter/Paramedic Gilbert died from complications due to a vehicle accident while transporting a patient to the hospital.

11/9/1996 John Bryant, Firefighter Volunteer, Age 21
Upper Gwynedd Township Fire Department, PA MVA

Firefighter Bryant was fatally injured in a motor vehicle accident while responding to an alarm. The firefighter was on his way to the station when his vehicle was hit from behind at high rate of speed.

11/9/1996 Steve Trice, Firefighter Volunteer, Age 24
Sharptown Volunteer Fire Department, MD Struck by Vehicle

Firefighter Trice stopped at a motor vehicle accident and was struck by a passing vehicle while attempting to extricate a victim.

11/12/1996 Walter Schwinger, Jr., Captain Career, Age 52
Tonawanda Fire Department, NY Pulmonary Embolism

Captain Schwinger died of a pulmonary embolism while asleep in the bunkroom while on duty.

11/12/1996 William Chambers, Firefighter Career, Age 49
Anne Arundel County Fire Department, MD Heart Attack

Firefighter Chambers collapsed and died of a heart attack during response to medical call.

| 11/24/1996 | Donald Manuel, Firefighter | Volunteer, Age 65 |
| | Highview Fire District, KY | Heart Attack |

Firefighter Manuel suffered a fatal heart attack upon arrival at the scene of a church fire.

| 11/27/1996 | Edward Ramos, Firefighter | Volunteer, Age 36 |
| | Branford Fire Department, CT | Trapped |

Firefighter Ramos was killed in a warehouse fire at Floors and More, Inc., after the roof collapsed, trapping him and two other firefighters inside. Despite having his SCBA facepiece knocked off in the collapse, Firefighter Ramos stayed on the hoseline and knocked down the fire so the other firefighters could escape.

| 12/4/1996 | Ruben Lopez, District Chief | Volunteer, Age 39 |
| | Houston Volunteer Fire Department, TX | Trapped |

District Chief Lopez was killed in a residential structure fire while attempting to rescue one of the house's occupants. The firefighter and the victim were caught in a flashover. Both the firefighter and the victim were killed.

| 12/8/1996 | Craig Arnone, Firefighter | Volunteer, Age 23 |
| | Somers Fire Department, CT | Electrocution |

Firefighter Arnone was electrocuted when his SCBA tank came into contact with a downed power line carrying 23,000 volts at a residential structure fire. A snowstorm was responsible for the power line being down. Firefighters thought the electrical power to the area had been shut off when they came to the house.

| 12/21/1996 | Stanley Scott, Firefighter | Career, Age 45 |
| | Chicago Fire Department, IL | Heart Attack |

Firefighter Scott suffered a fatal heart attack after hooking up to a hydrant at a structure fire. CPR was initiated on the scene, but firefighter Scott was pronounced dead at a hospital.

| 12/21/1996 | James A. Ellis, Firefighter | Career, Age 43 |
| | Boston Fire Department, MA | Trauma |

Firefighter Ellis died as a result of injuries sustained after falling approximately 20 feet down a fire pole on the way to a call. The presence of water possibly from a sink is listed as the cause of the fall. The fall caused severe head trauma and neurological damage.

| 12/21/1996 | Laura Halsey, Firefighter | Volunteer, Age 23 |
| | Stroh Volunteer Fire Department, IN | MVA |

Firefighter Halsey was driving to the hospital with a patient from an automobile wreck when a car struck them head on at 4:30 a.m. The striking car had no headlights and was in the wrong lane. All of the car's occupants were killed.

| 12/23/1996 | James Warick, Firefighter | Volunteer, Age 52 |
| | Burnet Volunteer Fire Department, TX | Struck by Vehicle |

Firefighter James Warick was struck by a vehicle while directing traffic at an incident.

| 12/27/1996 | Charles Brant Chesney, Firefighter | Volunteer, Age 36 |
| | Forsyth County Fire Department, GA | Trauma |

Firefighter Chesney was killed while advancing a hoseline to the upper floor of a three-story condominium fire when the roof collapsed due to unseen fire spread. The two firefighters with him were able to escape.

| 12/29/1996 | Raymond Emmrich, Chief | Volunteer, Age 54 |
| | Kimball Township Fire Department, WI | Heart Attack |

Chief Emmrich suffered a heart attack while driving a pumper to a dwelling fire. After suffering the heart attack, the pumper ended up in a snow bank.

| 1/1/1997 | Brian D. Myers, Sr., Engineer | Volunteer, Age 47 |
| | Schuylerville Hose Company, NY | Burns |

Firefighter Myers and three other firefighters were operating the nozzle at the scene of a restaurant fire early on New Year's Day when the ceiling collapsed and a flashover occurred. Two of the firefighters were able to escape. A rescue team placed a nozzle through the front window to cool down the area where firefighters were trapped. They located and removed one of the trapped firefighters. They then reentered the building and found Firefighter Myers. He too was removed and emergency medical care was provided. Firefighter Myers went into cardiac arrest while en route to the hospital. He had suffered burns to 70 percent of his body. His son and one other firefighter were also injured.

| 1/2/1997 | David P. Janora, Assistant Chief | Volunteer, Age 49 |
| | Clarence Center Fire Department, NY | Cardiac |

At 9:30 p.m., Chief Janora went into cardiac arrest during a meeting at the fire station. Chief Janora had attended a vehicle fire earlier during the day.

| 1/2/1997 | Harold "Mac" E. McGowan, Firefighter/Safety | Volunteer, Age 70 |
| | OFficer, Union Fire Company #1, PA | Cardiac |

Following a structure fire, Firefighter McGowan removed his gear and went into cardiac arrest.

| 1/3/1997 | Arthur R. Ebert, Firefighter | Volunteer, Age 63 |
| | Fort Morrow Fire Department, OH | Heart Attack |

Firefighter Ebert responded to a structure fire and returned to the station to clean up when he was dispatched to a report of a house fire. Firefighter Ebert and another firefighter took an engine to the scene. They were the first to arrive. The second firefighter pulled a handline but was not getting any water. He went to investigate and found Firefighter Ebert had collapsed due to a heart attack.

| 1/7/1997 | Carl L. Ayers, Fire Police | Volunteer, Age 67 |
| | Newton–Ransom Fire Company, PA | Struck by Vehicle |

Fire Police Officer Ayers was struck by a car and killed while directing traffic at a motor vehicle collision.

| 1/8/1997 | H. Robert Hathaway, Chief | Volunteer, Age 58 |
| | Branchport Fire Department, NY | Heart Attack |

Chief Hathaway collapsed due to a heart attack immediately following a meeting at the firehouse.

| 1/10/1997 | Harold Hester, Firefighter | Volunteer, Age 52 |
| | Malden Fire Department, MO | MVA |

Firefighter Hester was involved in a vehicle accident while responding to a call. He was in his personal vehicle and lost control when the car hit a patch of ice on the highway.

| 1/10/1997 | Allen H. Martin, Jr., Firefighter | Career, Age 33 |
| | New Orleans Fire Department, LA | Trapped |

Firefighter Martin became trapped by debris in a two-story residential structure fire while searching the structure for victims and conducting an interior attack. The roof collapsed and it was several minutes before firefighters were able to make a rescue attempt. Firefighter Martin died shortly after arriving at the hospital.

| 1/14/1997 | Stoy Geary, Chief | Volunteer, Age 62 |
| | Rosine Volunteer Fire Department, KY | Cardiac |

Chief Geary suffered cardiac arrest at the scene of a residential structure fire.

| 1/15/1997 | Richard Sanders, Lieutenant | Career, Age 47 |
| | Oakland Fire Department, CA | Heart Attack |

Lieutenant Sanders died as a result of a heart attack that occurred early in the morning in the station towards the end of his shift.

| 1/20/1997 | Thomas C. Reynolds, Firefighter | Volunteer, Age 29 |
| | Terlton Community Fire Department, OK | Burns |

Firefighter Reynolds was seriously burned while rescuing two children from an overturned dune buggy. He died 1/20/1998.

| 1/26/1997 | Robert William Martinson, Sr., Assistant Chief | Volunteer, Age 43 |
| | Conover Fire Department, WI | Smoke Inhalation |

Chief Martinson was on the roof of a house checking the progress of the fire when the roof collapsed.

| 2/4/1997 | Wayne M. Fogel, Firefighter | Career |
| | Detroit Fire Department, MI | Heart Attack |

Firefighter Fogel suffered from a heart attack while on duty at the firehouse.

| 2/5/1997 | Kevin C. Seaburg, Assistant Chief | Volunteer, Age 38 |
| | Selkirk Fire District, NY | Heart Attack |

Chief Seaburg collapsed due to a heart attack while carrying and setting up equipment at the scene of a structure fire.

2/6/1997	Bryan J. Golden, Firefighter	Age 21, Career
	Brett A. Laws, Firefighter	Age 29, Career
	Stockton Fire Department, CA	Trapped

Units were dispatched to a report of a house fire. The first arriving officer found a working fire and immediately requested a second-alarm assignment. Two houses were on fire and there was a possibility of a person trapped. Unbeknownst to the initial crews, the house was much bigger than it appeared from the street and there was a large two-story addition heavily involved in fire. An interior attack was initiated with a 1-3/4-inch handline through the front door.

Approximately 21 minutes later, with no warning, there was a catastrophic collapse of the entire second floor and roof of the addition. The collapse trapped firefighters working on the first floor. Fire Captain Oscar Barrera was trapped in the burning debris, but was rescued through the efforts of other firefighters. Captain Barrera was seriously burned. Firefighters Laws and Golden could not be rescued and were killed. This was Firefighter Golden's first fire.

The owner of the house was also killed in the fire. The second story had been added on by the owner for use as a dance studio and was made of heavy timber.

| 2/15/1997 | Timothy J. Warren, III, Firefighter | Volunteer, Age 36 |
| | Geneva Fire Department, NY | Heart Attack |

Firefighter Warren collapsed due to a heart attack while fighting a fire that broke out in a three-story dormitory at Hobart College. The fire started on the first floor. No students were injured.

| 2/16/1997 | Peter Kahn, Firefighter | Volunteer, Age 75 |
| | Trumansburg Fire Department, NY | Heart Attack |

Firefighter Kahn collapsed due to a heart attack while directing traffic at the scene of a two-car vehicle collision.

| 2/17/1997 | Charles "Chuck" H. Williams, II, Firefighter | Career, Age 29 |
| | Lexington Fire Department, KY | Burns |

Firefighter Williams and a second firefighter became trapped after entering a residential fire and falling through a hole into the basement. Both received second- and third-degree burns. Efforts were made to revive Firefighter Williams on the scene. The other firefighter was admitted to the hospital with serious burns.

2/22/97   **Robert E. Fowler, Firefighter**   **Volunteer, Age 54**
             Spencerport Fire Department, NY   **Trauma**

Firefighter Fowler was crushed and killed when a tree fell on his personal vehicle during a response to an emergency call. Due to extreme weather, the fire department had been called for several downed power lines and windows that had been blown out. Firefighter Fowler and his son were driving the half mile from their house to the station. His son, who was a junior firefighter, was also taken to the hospital with back injuries.

2/28/1997   **Charles Allen Weber, Sr., President**   **Volunteer, Age 48**
             Violetville Volunteer Fire Company, Baltimore   **Heart Attack**
               County, MD

Firefighter Weber returned from an emergency call and was in the process of storing his gear when he suffered a fatal heart attack.

3/15/1997   **Russett "Rusty" S. Hauber, Firefighter/Dive Team**   **Career, Age 32**
             Charlie Mestaz, Firefighter/Dive Team   **Volunteer, Age 36**
             Yakima County Sheriff's Dive Team, WA   **Asphyxiation**

Firefighters Hauber and Mestaz died from asphyxiation while attempting to rescue two civilian divers from a 1,000-foot long, 100-foot deep irrigation siphon. A police officer was also killed at this incident.

3/21/1997   **Tommy T. Gross, III, Firefighter**   **Career, Age 24**
             Tuscaloosa Fire Department, AL   **Heart Attack**

Firefighter Gross suffered from a heart attack while going through a burn building while in rookie school.

4/17/1997   **Larry L. Mercer, Captain**   **Career, Age 48**
             Duncan Fire Department, OK   **Stroke**

Captain Mercer died as a result of a stroke that occurred while on duty. He was on a 24-hour shift and was feeling ill. This prompted him to make a doctor's appointment for the next working day. He died as a result of the stroke 48 hours after experiencing pain on his way to work.

4/20/1997   **William W. Babka, Firefighter/Co-Pilot**   **Career, Age 34**
             Walter John Hirth, Jr., Captain/Pilot   **Career, Age 45**
             Pennsylvania Bureau of Forestry, Harrisburg, PA   **Aircraft Crash**

Pilot John Hirth and Co-Pilot William W. Babka were killed after approaching a forest fire to make a water drop. After leveling for the drop, the aircraft was affected by a downdraft from wind gusts. The smoke from the fire also had an impact on their visibility. The aircraft stripped off the tops of trees for approximately 100 feet before coming to rest. An unattended campfire started the blaze.

4/26/1997   **Earl Holsapple, Captain**   **Volunteer, Age 45**
             California Division of Forestry, La Cima Fire Center, **Asthma Attack**
               Julian, CA

Captain Holsapple suffered a severe asthma attack and died while teaching an equipment operating class at the California Division of Forestry's Fire Academy.

5/3/1997   **Jessie F. Bricker, Jr., Fire Apparatus Operator**   **Career, Age 47**
             San Antonio Fire Department, TX   **Cardiac**

Fire Apparatus Operator Bricker died as a result of cardiac arrest while fighting a four-alarm motel fire. He experienced smoke inhalation at the fire and, after returning to the station, FAO Bricker complained of feeling sick. He was transported to the hospital at 5:56 a.m. in severe respiratory distress. He died at 4:38 p.m. Other personnel also suffered from various symptoms as a result of exposure to the smoke. The following day, an environmental consultant was directed to sample and analyze the scene and clothing worn by firefighters. Analysis revealed that an "unusual chemical event" occurred at the scene of the fire. Testing determined firefighters may have been exposed to chlorine or chlorine compounds, hydrochloric acid, pesticides, amines, illicit drugs (such as methamphetamines), and other undetermined chemicals.

5/3/1997	**Tracy D. Floyd, Firefighter**	**Career, Age 29**
	Winchester Fire Department, TN	**MVA**

Firefighter Floyd was killed while responding to the scene of a structure fire when another vehicle pulled out in front of him, causing a collision.

5/5/1997	**Timothy M. Goff, Firefighter**	**Volunteer, Age 27**
	Kenmore Volunteer Fire Department, NY	**Trauma**

Firefighter Goff died on 5/24/1997, as a result of injuries sustained from a wall collapse during a paint store fire. Five other firefighters were injured in the collapse.

5/8/1997	**M. Edward Hudson, Lieutenant**	**Career, Age 53**
	Reginald G. Robinson, Sr., Firefighter	**Volunteer, Age 33**
	Stewart Warren, Captain	**Career, Age 47**
	West Helena Fire Department, AR	**Explosion**

The West Helena Fire Department was dispatched to the BPS Bartlo chemical plant at 1:02 p.m. for a report of smoke coming from the building. Firefighters were advised that the building contained azinphos methyl. The building exploded at approximately 1:22 p.m., killing three firefighters and severely injuring one. Eleven other firefighters were involved in the rescue of the injured firefighter and the rescue attempt for the three firefighters who died.

5/9/1997	**Will Ellis Rowe, Jr., Captain**	**Career, Age 49**
	Macon–Bibb County Fire Department, Macon, GA	**Trauma**

During a severe thunderstorm, Captain Rowe and his crew responded to a report of a trash fire and found a downed tree on the roof of a house. Captain Rowe and another firefighter went behind the house to assess the situation when the tree began to slide off the roof. The firefighters began to move out of the way when Captain Rowe slipped in the mud and was crushed by the falling tree. He was killed instantly.

5/12/1997	**Lawrence Hobson, Lieutenant**	**Career, Age 49**
	Robbins Fire Department, IL	**Heart Attack**

Lieutenant Hobson collapsed and died due to a heart attack while pulling 2-1/2-inch hoseline at an abandoned house fire.

5/25/1997	**William "Junior" T. Wilson, Firefighter**	**Volunteer, Age 41**
	Pinecrest Volunteer Fire Department, Jacksboro, TN	**Apparatus Rollover**

Firefighter Wilson was killed during a rollover en route to a vehicle accident on Interstate 75.

5/26/1997	**Stanley F. Kaminski, Firefighter**	**Volunteer, Age 67**
	Langford–New Oregon Fire Department,	**Cardiac**
	North Collins, NY	

Firefighter Kaminski went into cardiac arrest and died while attending a memorial service with the fire department.

5/29/1997	**David M. Ray, Firefighter**	**Contract, Age 21**
	California Department of Forestry/Conservation	**Heat Stroke**
	Corps, Julian, CA	

Firefighter Ray died as a result of a heat stroke after fighting a brush fire. The fire started when a tractor mower hit a rock, producing a spark. Over 140 firefighters fought the fire. Two firefighters went to the hospital after collapsing from heat stroke. According to news articles, this was Firefighter Ray's first fire. Firefighter Ray's crew was assigned to cut a fireline on a hillside.

Winds at the time were light and the temperature was close to 100 degrees by early afternoon. The humidity level was under 15 percent. After a long period cutting firelines (which included one break), Firefighter Ray manifested symptoms of heat stress. Despite the immediate on-scene attention of his crew and quick evacuation to a hospital, his condition deteriorated rapidly to heat stroke. He did not regain consciousness and died early the next morning.

6/3/1997	**Jesse Gates, Pilot**	**Career, Age Unknown**
	Leo A. Stevens, Firefighter	**Career, Age 55**
	Fort Apache Indian Reservation and BIA Facility Management, AZ	**Aircraft Crash**

Pilot Jesse Gates and Firefighter Leo Stevens were killed in the crash of a fire reconnaissance airplane on the Fort Apache Indian Reservation in Eastern Arizona. A lookout tower reported black smoke near the Black River. Dispatch lost contact with the patrol plane at about the same time the smoke was reported. The plane had been on a routine fire reconnaissance flight.

6/5/1997	**James H. Johnson, Firefighter**	**Volunteer, Age 63**
	Charles A. Rudd, Lieutenant	**Volunteer, Age 21**
	New Bloomington Fire Department, OH	
	Robert Douglas Good, Firefighter/Paramedic	**Career, Age 30**
	Rural Metro Ambulance	**Electrocution**

Firefighter Johnson and Lieutenant Rudd were electrocuted at the scene of a motor vehicle collision when a rescuer came into contact with downed power lines, creating a chain reaction. Five rescuers were injured and three rescuers were electrocuted. The initial victim, who was being carried on a backboard, was also fatally electrocuted.

6/7/1997	**Gerald T. Ertle, Captain**	**Volunteer, Age 53**
	Benton Volunteer Fire Department, MI	**Heart Attack**

Captain Ertle went into cardiac arrest and died during a training class at the state academy. After completing certification classes during April, May, and June, he attended the volunteer certification field day (practical testing day) at the fire academy. Shortly after completing the testing, Captain Ertle suffered a fatal heart attack.

6/9/1997	**Timothy Wayne Martin, Firefighter/EMT**	**Career, Age 38**
	Clovis Fire Department, NM	**Apparatus Rollover**

Firefighter Martin was providing patient care in the back of an ambulance when the ambulance lost control on wet pavement and overturned. The patient was also killed.

6/16/1997	**Edwin J. Haungs, Sr., Fire Police**	**Volunteer, Age 51**
	South Lockport Fire Company, Inc., NY	**Heart Attack**

Fire Police Officer Haungs' fire company was called to assist with traffic control at a mutual-aid motor vehicle accident. He collapsed and died due to a heart attack immediately after returning from this call.

6/16/1997	**William Jack Northam, Firefighter**	**Volunteer, Age 55**
	Laurel Fire Department, Inc., DE	**Heart Attack**

The Laurel Fire Department was dispatched to a motor vehicle accident. Firefighter Northam was putting away tools before boarding the apparatus when he collapsed due to a heart attack. He was taken to the hospital and died 10 hours later.

6/16/1997	**John McClay Watson, Firefighter**	**Volunteer, Age 18**
	Moscow Volunteer Fire Company, PA	**MVA**

Firefighter Watson died as a result of injuries sustained from an accident involving his personal vehicle while en route to an EMS call. He was thrown from his vehicle, which rolled over three times in the course of the accident. He died the next day from severe head trauma.

6/19/1997 William C. Mellon, President Volunteer, Age 58
Bay Ridge Volunteer Fire Department, Heart Attack
Lake George, NY

Firefighter Mellon suffered from a heart attack while preparing to respond in his personal vehicle to a fire alarm.

6/20/1997 Michael F. Drobitsch, Firefighter Career, Age 46
Chicago Fire Department, IL Dive Accident

Firefighter Drobitsch's death occurred during a diving training session.

6/22/1997 Michael E. Neuner, Sr., Lieutenant Volunteer, Age 35
Brewster Fire Department, NY Trapped

Lieutenant Neuner died due to injuries sustained at a residential structure fire after becoming trapped in the basement.

6/28/1997 Ricky G. Moore, Firefighter Volunteer, Age 29
Oak Grove/Thach Fire Department, AL Apparatus Rollover

Firefighter Moore was killed when he was thrown from a fire truck after it overturned en route to a fire call.

7/1/1997 Joseph M. Vagnier, Firefighter/EMT Volunteer, Age 21
Monroeville Volunteer Fire Company #4, PA Drowning

Firefighter Vagnier and two other firefighters were attempting to rescue a flood victim when Firefighter Vagnier was swept under the wheel of a truck by raging waters. Firefighters treated him at the scene, and immediately transported him to a local hospital where he subsequently died. Reports indicate that either Firefighter Vagnier was pulled under when another firefighter holding his security rope went under or he was pulled under after attempting to help rescue another firefighter that went down. He died on 7/3/97.

7/4/1997 Michael L. Seguin, Firefighter Career, Age 31
Buffalo Fire Department, NY Trapped

Firefighter Sequin was killed when he became trapped by a roof collapse while fighting a residential structure fire. One other firefighter was injured and suffered second-degree burns. The second firefighter was dragged to safety after becoming unconscious. Rescuers did not see Firefighter Sequin due to heavy smoke, and he was not located until later that afternoon. Fire officials stated that there was a possibility that fireworks started the fire. The owner of the house believed that a "rocket" landed on her roof.

7/6/1997 Floyd Dean Hiser, Sr., Pilot Career, Age 51
Sierra National Forest, Clovis, CA Helicopter Crash

Pilot Hiser was killed as a result of a helicopter crash that occurred during a water drop at a wildland fire.

7/13/1997 James H. Tebo, Captain Volunteer, Age 61
Ranker Community Fire Department, Bangor, MI Heart Attack

Captain Tebo collapsed at the scene of a structure fire and later died as a result of a heart attack.

7/15/1997 Malcolm A. Rovero, Firefighter Career, Age 34
Estero Fire Protection and Rescue Service District, FL Cardiac

Firefighter Rovero died from apparent cardiac and respiratory arrest at the scene of a 10-acre brush fire.

7/16/1997 Albert Sippel, Lieutenant Volunteer, Age 50
Bellmore Fire Department, NY Seizure

Lieutenant Sippel collapsed of a seizure while assisting a victim at a motor vehicle accident.

7/25/1997 Jerome H. Chlian, Jr., Firefighter   Volunteer, Age 46
            Starbuck Fire Department, MN   Heart Attack

Firefighter Chlian died as a result of a heart attack while on duty.

8/3/1997   Joseph J. Estavillo, Fire Engineer   Career, Age 44
            San Diego Fire & Life Safety Services, CA   Infectious Disease

Fire Engineer Estavillo was called out on a strike team to fight a brush fire in the northern part of San Diego County. While fighting the fire, he sustained cuts on his hands (through his gloves). After returning to his crew that night (2:00 a.m.), Estavillo complained to his Captain about not feeling well. They immediately went to the hospital. By morning they had discovered that Fire Engineer Estavillo was infected with Strep A, which caused his death.

8/19/1997 Jeffrey E. Sammons, Firefighter   Volunteer, Age 21
            South Whitley Fire Department, IN   Trauma

Firefighter Sammons was killed and two others were injured in a restaurant fire caused by cooking equipment that had been left on. Firefighter Sammons and others were making an internal fire attack when the heat buildup became extreme. They started to exit the structure when a flashover occurred causing some of the ceiling tile to fall.

8/22/1997 Richard B. Jenkins, Sr., Firefighter   Volunteer, Age 39
            Tennville Fire Department, GA   MVA

Firefighter Jenkins was killed en route to a house fire from his home in a private vehicle.

8/31/1997 Robert D. Chisholm, Assistant Chief   Volunteer, Age 50
            Gearhart Fire Department, OR   Heart Attack

Chief Chisholm had a heart attack while trying to rescue a missing swimmer at Gearhart beach in Oregon. After searching for the victim, Chief Chisholm became tired and passed out. His fellow firefighters dragged him to the shore. Crews were unable to revive him.

9/5/1997   Kenneth E. Bayer, Captain   Career, Age 52
            Los Angeles County Fire Department, CA   Carbon Monoxide

Captain Bayer died from cardiac arrest after being exposed to high concentrations of carbon monoxide (CO). The exposure to smoke and CO occurred over an approximate 45–60-minute period as he directed the interior extinguishment, salvage, and overhaul operations of a chimney and attic fire in a two-story condominium. He died 9/9/1997.

9/7/1997   Howard E. Strube, Firefighter   Career, Age 34
            Canton Fire Department, IL   Trauma

Firefighter Strube was killed during training on the fire department's new aerial apparatus. Firefighter Strube was operating as the safety observer (platform operation) when his head became caught between the ladder rungs while the ladder was retracted.

9/8/1997   David E. Carpenter, Firefighter   Volunteer, Age 38
            Donald J. Payton, Sr., Captain   Volunteer, Age 57
            Thayer Rural Fire Department, MO   MVA

Captain Payton and Firefighter Carpenter were killed while responding to a motor vehicle accident. The fire truck they occupied collided head on with a dump truck. The police reported that the dump truck crossed the centerline and struck the fire truck.

9/14/1997 Henry E. Perry, Firefighter   Volunteer, Age 59
            Pumpkin Center Fire Department, Inc.,   Trauma
            Jacksonville, NC

Firefighter Perry went into cardiac arrest at the scene of a structure fire. He was climbing on top of a fire truck when he fell. The pump operator noticed him falling, but had no time to react. Firefighter Perry died as a result of injuries received by the fall.

10/2/1997	Walter Douglas Buckert, Firefighter	Volunteer, Age 23
	Michael D. Mapes, Firefighter	Volunteer, Age 35
	Carthage Fire Department, IL	Explosion

These fatalities occurred at a fire located at a grain dryer fire north of Burnside, IL. En route to the fire, firefighters were advised that liquid propane gas tanks were involved. Upon arrival, firefighters noticed one of the 1,000-gallon tanks venting and shooting flames approximately 45 to 50 feet in the air. After surveying the scene and talking to the owner, firefighters decided to move the truck to a safer location for a better point of attack. While doing so, one of the tanks exploded, causing the deaths of Firefighters Mapes and Buckert. Two other firefighters were injured.

| 10/10/1997 | William "Pops" H. Winters, Deputy Chief | Volunteer, Age 76 |
| | Atglen Fire Company, PA | Cardiac |

Chief Winters collapsed and went into cardiac arrest at the scene of a structure fire.

| 10/15/1997 | Harold "Ray" Elliott, Battalion Chief | Career, Age 54 |
| | Kern County Fire Department, CA | Heart Attack |

Chief Elliott was doing mandatory physical training outside at Virginia Colony Station 41 when he collapsed due to a heart attack at approximately 9:40 p.m. He was on an overtime shift and had already worked approximately 24 hours. Chief Elliott died 4/28/1998.

| 10/22/1997 | David Shawn Williams, Firefighter | Career, Age 26 |
| | Taylor County Fire Rescue Service, FL | Shot (Accidental) |

Firefighter Williams was killed when an overheated bullet (0.22 caliber) from a greenhouse fire discharged, hitting him in his chest.

| 10/24/1997 | John M. Carter, Sergeant | Career, Age 38 |
| | District of Columbia Fire Department, Washington, DC | Trapped |

Sergeant Carter died when the floor beneath him collapsed during a three-alarm grocery store fire. The incident commander was evacuating the building at the time of the collapse. The investigation into the fire's cause indicated that faulty electrical wiring in the basement started the fire.

| 10/24/1997 | David Womer, Firefighter/EMT | Volunteer, Age 24 |
| | Mount Carmel Volunteer Fire Department Station 5, PA | Explosion |

Firefighter Womer was killed when the rescue squad building where he was on duty experienced an explosion. A fellow rescue squad member rolled an open 30-pound propane tank from a gas grill into the squad house as a practical joke. All but Firefighter Womer evacuated the building. Eventually the gas reached the pilot light in the furnace room and triggered the explosion.

| 10/25/1997 | Kathryn A. Mayfield, Firefighter/EMT | Volunteer, Age 47 |
| | Crooksville Volunteer Fire Department, OH | Heart Attack |

Firefighter Mayfield collapsed in the station after returning from a tire fire and was taken to the hospital. She died the next morning due to a heart attack.

10/27/1997	James E. Hynes, Firefighter	Career, Age 27
	Terry McElveen, Lieutenant	Career, Age 43
	Philadelphia Fire Department, PA	Smoke Inhalation

Lieutenant McElveen and Firefighter Hynes died as a result of smoke inhalation at the scene of a residential structure fire. The fire was a result of wires that had come down on the roof during a heavy rainstorm. The firefighters were operating in the interior of a two-story occupied dwelling with a fire in the basement. They ran out of air, removed their SCBA masks, and remained inside the dwelling. The two firefighters were found near the back door with their SCBAs on, but their masks off.

| 11/2/1997 | **Leroy Swenson, Captain** | Career, Age 56 |
| | **Minneapolis Fire Department, MN** | **Struck by Vehicle** |

Captain Swenson was killed at the scene of a four-vehicle accident when a large commercial truck lost control on the icy road and rolled over on top of him, killing him instantly.

| 11/5/1997 | **William S. Bradner, III, Firefighter** | **Volunteer, Age 30** |
| | **Tunstall Fire and Rescue Company, Danville, VA** | **Apparatus Rollover** |

Firefighter Bradner was killed when he was thrown from a tanker truck as it overturned while returning to the scene of a structure fire. Another volunteer was also injured in the accident and suffered a broken pelvis. The two firefighters had already delivered one load of water and were returning with the second load when the accident occurred.

11/6/1997	**Johnson "Jack" Oatman, Firefighter**	**Volunteer, Age 55**
	Ewansville Volunteer Fire Department,	**Heart Attack**
	Mount Holly, NJ	

Firefighter Oatman died from a heart attack that occurred while preparing to respond to a motor vehicle accident with entrapment.

| 11/7/1997 | **John F. Kroening, Firefighter** | **Volunteer, Age 75** |
| | **Cambria Volunteer Fire Company, Lockport, NY** | **Struck by Vehicle** |

Firefighter Kroening was on the scene of a motor vehicle accident setting up equipment when a passing vehicle struck him. He was taken to the hospital and died from the injuries the next day.

| 11/13/1997 | **Eugene Ottonello, Assistant Fire Manager Officer** | **Wildland Career, Age 47** |
| | **Bureau of Land Management, Battle Mountain, NV** | **Asthma Attack** |

Officer Ottonello died due to an asthma attack at a prescribed burn. He then went into respiratory and cardiac arrest and died.

| 11/14/1997 | **Scott Alan Vrabel, Firefighter** | **Volunteer, Age 26** |
| | **New Salem Volunteer Fire Department, PA** | **Apparatus Rollover** |

Firefighter Vrabel was killed when he lost control of a brush truck en route to a two-vehicle accident. The truck ran off the road into a telephone pole and flipped, pinning Firefighter Vrabel underneath.

| 11/15/1997 | **William H. Fairweather, Fire Police** | **Volunteer, Age 78** |
| | **Middletown Fire Department, NY** | **Heart Attack** |

Fire Police Officer Fairweather was directing traffic at the scene of a motor vehicle accident when he had a heart attack and died.

| 11/18/1997 | **George A. Davis, Firefighter 1st Class** | **Career, Age 27** |
| | **Houma Fire Department, LA** | **Heart Attack** |

Firefighter Davis entered a structure fire wearing breathing apparatus. He later exited the structure, still wearing SCBA, and collapsed and fell into the arms of another firefighter. Firefighters performed CPR on the scene. Davis was then taken to the hospital where he died on 11/23/97.

| 11/22/1997 | **Gregory I. Quinn, Assistant Chief** | **Volunteer, Age 46** |
| | **Village of Westfield Fire Department, WI** | **Struck by Vehicle** |

Chief Quinn died after being hit by a sport utility vehicle that lost control on an icy bridge at the scene of a motor vehicle accident. The car struck two firefighters and hit one of the cars involved in the first wreck. The other firefighter was not seriously injured. The driver of the SUV was not wearing her seat belt and was killed.

| 11/26/1997 | **George Hopey, Jr., Fire Police** | **Volunteer, Age 69** |
| | **Dravosburg Volunteer Fire Company #1, PA** | **Heart Attack** |

Fire Police Officer Hopey suffered a heart attack while directing traffic at the scene of a structure fire. He died on 11/28/97.

12/1/1997   **Thomas M. McCormack, Chief**   Career, Age 44
                    Watervliet Fire Department, NY   Heart Attack

Chief McCormack died from a heart attack while directing fire operations at a residential structure fire mutual-aid call in the City of Troy, NY. The City of Watervliet was first due on the call, and Troy units were deployed on the second alarm.

12/5/1997   **John F. Lincoln, Jr., Firefighter**   Volunteer, Age 52
                    Purcellville Volunteer Fire Department, VA   Heart Attack

Firefighter Lincoln suffered a heart attack after returning from an all night working residential fire that included a civilian casualty. He died on 12/6/97.

12/9/1997   **"Randy" Smartt, Firefighter**   Career, Age 49
                    Huntsville Fire Department, AL   Heart Attack

Firefighter Smartt died as a result of a heart attack that occurred while on duty at the firehouse.

12/11/1997   **Ronald A. Guilmette, Private**   Career, Age 38
                    Woonsocket Fire Department, RI   Unknown

Private Guilmette responded to a fire alarm at 1:30 a.m. Upon returning to the station at 1:40 a.m., Private Guilmette complained of back pain to his crew and told them he thought he would be more comfortable sitting up on the couch in the television room. At 5:30 a.m., his engine was dispatched to another alarm. The crew found Private Guilmette collapsed on the floor near the couch.

12/15/1997   **Leonard N. Zeller, Firefighter**   Volunteer, Age 53
                    Edwards Fire Department, NY   Heart Attack

Firefighter Zeller died as a result of a heart attack that occurred while responding to an EMS call. He was on foot on the way to the fire station when he collapsed. He was revived on the way to the hospital, but died at the hospital.

12/17/1997   **Scott M. Berry, Firefighter/Driver**   Volunteer, Age 33
                    Bradley County Volunteer Fire Department,   Apparatus Rollover
                    Cleveland, TN

Firefighter/Driver Berry was killed when his 12,050-gallon tanker overturned while responding to the scene of a brush fire. His brother was also in the tanker and was transported by helicopter to the hospital. Firefighter Berry was wearing his seat belt at the time of the accident.

12/20/1997   **William "Sam" Smitherman, Sr., Firefighter**   Volunteer, Age 59
                    East Oktibbeha Fire Department, Starkville, MS   Struck by Vehicle

A passing car at a motor vehicle fire struck and killed Firefighter Smitherman as he was retrieving a crowbar from the engine.

12/23/1997   **Brian T. Hauk, Assistant Chief**   Volunteer, Age 32
                    Logan–Trivoli Fire Department, Hanna City, IL   MVA

Chief Hauk died in a vehicle accident while responding to the firehouse for a reported oven fire in an apartment complex. Chief Hauk was taking evasive action to avoid another vehicle that failed to yield to his vehicle when his vehicle flipped. He was displaying an activated blue light. He had the right of way and the other car had a stop sign.

12/30/1997   **Thomas P. Ryan, Firefighter**   Volunteer, Age 68
                    Middletown Township Fire Department, NJ   Heart Attack

Firefighter Ryan died of a heart attack after returning from a call.

1/5/1998   **Harold E. Roemer, Jr., Firefighter**   Volunteer, Age 55
                    Greenlawn Fire District, NY   Heart Attack

Firefighter Roemer had just completed 30 minutes on a treadmill in the gym located at fire department headquarters. He signed out of the gym, drank some water, and collapsed due to a heart attack.

1/6/1998   **Prince Albert Mousley, Jr., Firefighter**   **Career, Age 58**
                            Wilmington Fire Department, DE   **Heart Attack**

Firefighter Mousley was a member of a ladder company on the scene of an oil burner fire in the basement of a residential structure. As he and a partner entered the rear of the structure, Firefighter Mousley stated that he was tired and collapsed of a heart attack. Firefighter Mousley had just returned to duty after a battle with cancer. Further information related to this incident can be found in NIOSH Fire Fighter Fatality Investigation 98–F–13.

1/12/1998   **Robert J. O'Toole, Firefighter**   **Part-Time Paid, Age 25**
                            Washington Township Fire Department, OH   **Struck by Vehicle**

Firefighter O'Toole responded to an automobile collision on an interstate highway. The victim of the original collision had been loaded into an ambulance and had left the scene. As Firefighter O'Toole and others began to disconnect the battery on the vehicle, which was located in the median, he was struck and killed by another vehicle that had lost control on the ice. A police officer was also killed and another firefighter was severely injured in this incident.

1/16/98   **Brian Allen Cannon, Training Officer**   **Volunteer, Age 30**
                            Taylors Bridge Fire Department, Inc., NC   **Apparatus Rollover**

Firefighter Cannon and a fire captain were returning to the station from the scene of a traffic collision. The pumper left the roadway and overturned. Firefighter Cannon, who was not wearing a seat belt, was ejected and sustained blunt trauma injuries to the head.

1/21/1998   **Gregory Scott Carter, Firefighter**   **Volunteer, Age 24**
                            Fairlea Volunteer Fire Department, WV   **Carbon Monoxide/Burns**

Firefighter Carter responded to a report of smoke in a supermarket. The market was contained in a strip mall, which also included a post office and a photoprocessing store. Firefighter Carter had been employed at the supermarket in the past. Firefighter Carter and a captain entered the front of the store in full protective clothing and SCBA to search for the fire. They became disoriented while trying to exit the store. The captain alerted other firefighters by radio that he and Firefighter Carter were lost and in need of rescue. Firefighter Carter ran out of air and placed the breathing tube from his SCBA into his coat in an attempt to breathe. The captain was able to escape without significant injury. Immediate attempts were made by on-scene firefighters to rescue Firefighter Carter but rescuers were driven back by intense heat and smoke. Firefighter Carter was wearing a PASS device but it was not turned on. No hose line or search rope was used. The cause of death was smoke and soot inhalation, carbon monoxide poisoning, and complete body charring. This was an accidental fire caused by an electrical malfunction in a wall. Further information related to this incident can be found in NIOSH Fire Fighter Fatality Investigation 98–F–04.

1/27/98   **Stephen Earl Murphy, Lieutenant**   **Career, Age 47**
                            Philadelphia Fire Department, PA   **Heart Attack**

Lieutenant Murphy responded to a fire in a row house dwelling. He carried a 16-foot portable ladder to the rear of the structure, raised the ladder, broke out windows, climbed the ladder, and entered a bedroom. Other firefighters who ascended the ladder reported seeing Lieutenant Murphy in the bedroom. He ordered them to proceed into the structure and continue ventilation. When the firefighters returned to the bedroom, they found Lieutenant Murphy face down and unresponsive suffering from an apparent heart attack. Firefighters initiated emergency medical treatment and Lieutenant Murphy was transported to the hospital where a heartbeat was restored. Lieutenant Murphy died on 2/3/1998. The cause of the fire was ruled accidental as a result of a portable kerosene heater placed too close to combustibles. Lieutenant Murphy was not wearing SCBA.

2/5/1998	Stephen D. Carletti, Firefighter	Volunteer, Age 43
	David P. Theisen, Firefighter	Volunteer, Age 29
	Crooksville Fire Department, OH	Asphyxiation/Burns, Trauma

Firefighters Carletti and Theisen responded to a report of a fire in the basement of a single-story home. They entered the basement with other firefighters and extinguished fire in the ceiling. In the process of moving around the basement, the attack line was pinched off when it was caught in a folding chair. Firefighters were not aware that their water supply had been cut off. When they began to pull additional ceiling tiles, the room experienced a flashover. Of the five fire-fighters in the basement when the flashover occurred, two escaped, one was rescued, and two were killed. An adjacent room, which had not been discovered by the firefighters, was fully involved in fire and fire spread to the other room when tiles were removed. Repeated radio requests for help and water were received from the basement but rescuers were unable to reach the firefighters in distress due to severe heat and fire. Both firefighters were wearing their PASS devices, they were turned on, and they activated. The fire cause was determined to be accidental. Firefighter Carletti died of asphyxiation and burns and Firefighter Theisen died as the result of a crush injury. Firefighter Theisen was also a career firefighter in Westerville. Further information related to this incident can be found in NIOSH Fire Fighter Fatality Investigation 98–F–06.

| 2/10/1998 | Richard L. Kalous, Firefighter | Career, Age 50 |
| | De Pere Fire Rescue, WI | Heart Attack |

Firefighter Kalous responded as a member of an engine company to a car fire. Upon arrival at the scene, he hand stretched a 5-inch supply line to a fire hydrant about 75 feet from the engine. When it was determined that the supply line would not be needed, he was directed to don SCBA and assist with fire attack. He was discovered by other firefighters on a side step of the engine unresponsive and suffering from a heart attack; firefighters were unable to revive him.

| 2/11/1998 | Warren D. Myers, Jr., Firefighter | Career, Age 48 |
| | Tulsa Fire Department, OK | Heart Attack |

Firefighter Myers and his crew responded to a gas leak at a single-family dwelling. The line was shut off and, after repairs were made, Firefighter Myers turned the gas back on. Firefighter Myers was observed to be fatigued at the incident. When his engine company returned to quarters, Myers did not get off the truck and appeared to be in distress. Despite immediate medical treatment by his crew and others, he died of a heart attack. Further information related to this incident can be found in NIOSH Fire Fighter Fatality Investigation 98–F–29.

2/11/1998	Patrick Joseph King, Firefighter/Paramedic	Career, Age 40
	Anthony E. Lockhart, Firefighter	Career, Age 40
	Chicago Fire Department, IL	Smoke Inhalation

Firefighter King and Firefighter Lockhart responded from different companies to a report of a structural fire in a tire shop. No visible fire was encountered, there was no excessive heat, and only light smoke was found in most of the building with heavier smoke in the shop area. Ten firefighters were in the interior of the structure when an event that has been described as a flash-over or backdraft occurred, disorienting the firefighters. Some were able to escape but Firefight-ers King and Lockhart were trapped in the structure. A garage door that self-operated due to fire exposure may have introduced oxygen into the fire area and may have been a factor in the backdraft. The exit efforts of firefighters were complicated by congestion in the building. Within minutes of the backdraft, the building was completely involved in fire and rescue efforts were impossible. Both firefighters died from carbon monoxide poisoning due to inhalation of smoke and soot. Further information related to this incident can be found in NIOSH Fire Fighter Fatality Investigation 98–F–05.

| 2/5/1998 | Luis A. Rivera–Rivas, Firefighter | Career, Age 58 |
| | Puerto Rico Fire Department, PR | Heart Attack |

Firefighter Rivera–Rivas died of a heart attack after being exposed to smoke at a grass fire in the town of Papillas. The wind shifted, resulting in the smoke exposure.

2/17/1998   **Keith C. Thomas, Fire Chief**   **Volunteer, Age 56**
                  Aubbeenaubbee Volunteer Fire Department, IN   **Heart Attack**

Chief Thomas was helping to prepare for a CPR training session in his fire station when he collapsed and died from a heart attack.

2/25/1998   **William E. Bonnar Sr., Battalion Chief**   **Career, Age 61**
                  Orland Fire Protection District, IL   **Heart Attack**

Chief Bonnar collapsed and died of a heart attack approximately 20–30 minutes after the completion of an SCBA drill in a commercial structure.

3/8/1998   **Joseph C. Dupee, Fire Captain I**   **Career, Age 38**
                  Los Angeles City Fire Department, CA   **Asphyxiation/Burns**

Captain Dupee and his company were dispatched to a structure fire in a pet food processing company and were assigned to backup interior crews. When fire conditions worsened, all firefighters exited the building with the exception of Captain Dupee who had somehow been separated from his crew. The situation was further complicated by the activation of an emergency signal by another firefighter that had become disoriented (he was rescued by his company officer). Shortly after firefighters left the building, a partial roof collapse occurred. When it was determined that Captain Dupee was missing, a rapid intervention crew forced entry in the rear of the structure and removed Captain Dupee. He was burned over 95 percent of his body. Rescuers initiated advanced life support care, but he was pronounced dead at the hospital. The cause of death was determined to be asphyxiation and burns. The fire was accidental and started as a grease fire in a convection oven. Further information related to this incident can be found in NIOSH Fire Fighter Fatality Investigation 98–F–07.

3/8/1998   **Edward J. Matter, Jr., Firefighter**   **Volunteer, Age 56**
                  Westford Volunteer Fire Department, NY   **Trauma**

Firefighter Matter and other members of his department were dispatched to a report of a downed tree blocking the roadway. Firefighter Matter and another fire department member arrived at the scene in their personal vehicles. Each carried a chain saw and began to remove the tree from the roadway. The firefighters agreed that they would use a rope to pull the remnants of the tree to the ground to make a safer operation. While the rope was being prepared, Firefighter Matter continued to remove loose debris and began to use his chain saw to cut at one of the larger branches supporting the tree. Firefighter Matter was caught and carried by the tree as it rotated and was pinned face down under the largest section of the tree. Despite immediate removal of the tree by other firefighters, Firefighter Matter died from crush injuries.

3/9/1998   **David John Good, Firefighter**   **Volunteer, Age 36**
                  Lionville Fire Company, PA   **Struck by Vehicle**

Firefighter Good was killed when he was struck by a tractor-trailer truck that had lost control and slid into firefighters providing treatment at the scene of an earlier automobile collision. Firefighter Good was in the rear of the ambulance when he was struck. Nine other responders were injured, three of them severely. All emergency response personnel were out of the travel lanes when the incident occurred. The incident occurred in heavy rain.

3/16/1998   **Paula Bennett, Firefighter**   **Volunteer, Age 44**
                  Carriere Volunteer Fire Department, MS   **Trauma**

Firefighter Bennett died in a motor vehicle accident while responding to a structure fire. She was the only female firefighter to die in the line of duty during 1998.

3/23/1998   **Michael A. Butler, Firefighter/Lead Paramedic**   **Career, Age 33**
                  Michael D. McComb, Apparatus Operator   **Career, Age 48**
                  Eric F. Reiner, Firefighter/Lead Paramedic   **Career, Age 33**
                  Los Angeles City Fire Department, CA   **Helicopter Crash**

Firefighters Butler and Reiner and Apparatus Operator McComb died when the fire department helicopter in which they were flying crashed in a park. The crash occurred while they were transporting an 11-year-old child that had been injured in a vehicle collision to the hospital. In addition to these fatalities, the child was killed in the crash and the pilot and one additional crewmember were severely injured. The reason for the crash has been attributed to the inflight failure of the tail rotor system. More information related to this incident can be found in National Transportation Safety Board report LAX98GA127.

| 3/28/1998 | **Richard K. Rice, Sr., Assistant Chief** | **Volunteer, Age 38** |
| | **Nassauville Volunteer Fire Department, FL** | **Apparatus Rollover** |

Chief Rice was killed in a vehicle collision while en route to the scene of an illegal burn. Rice was operating a pumper when the truck left the road, rolled one and a half times, and ended up on its roof in a ditch. Chief Rice was partially ejected and was pronounced dead at the scene.

| 4/1/1998 | **Jeffrey William Reick, Safety Officer** | **Volunteer, Age 34** |
| | **Aurora Fire Department, IN** | **Gastric Hemorrhage** |

Safety Officer Reick experienced a gastric hemorrhage while setting up for a live-fire training exercise. He was stricken when he exited the structure after another firefighter set the training fires. Safety Officer Reick died on 4/2/98.

4/9/1998	**Thomas James Archer, Jr., Firefighter**	**Volunteer, Age 46**
	Larry R. Walsh, Firefighter	**Volunteer, Age 45**
	Albert City Community Fire Department, IA	**Explosion**

Firefighter Archer and Firefighter Walsh were killed when they were struck by pieces of an 18,000-gallon propane tank when the tank experienced a BLEVE. The piping leading from the tank was damaged when an all-terrain vehicle struck it. A fire developed as a result of the leak and the fire department responded. While firefighters were protecting exposures, the tank exploded. Six other firefighters and a deputy sheriff were injured in the explosion. More information related to this incident is available in NIOSH Fire Fighter Fatality Investigation 98–F–14, report number 98–007–I–IA, from the U.S. Chemical Safety and Hazard Investigation Board and from the National Fire Protection Association.

| 4/13/1998 | **Michael Curtis Wiborg, Firefighter** | **Paid-On-Call, Age 46** |
| | **Chanhassen Fire Department, MN** | **Heart Attack** |

Firefighter Wiborg died of a heart attack after completing an annual physical agility test.

| 4/22/1998 | **Ralph William Stanbery, Firefighter** | **Volunteer, Age 62** |
| | **Granby Fire Department, MO** | **Heart Attack** |

Firefighter Stanbery was assisting with the filling of brush trucks from a tanker at an arson wildland fire. He collapsed and subsequently died of a heart attack.

| 4/25/1998 | **William J. Robertson, Battalion Chief** | **Career, Age 46** |
| | **Ridge Road Fire District, NY** | **Heart Attack** |

After exercising for 45 minutes on a treadmill at the fire station, Chief Robertson responded to a report of a car fire. When the incident proved to be in another jurisdiction, Chief Robertson began to return to quarters. He suffered a heart attack, his command vehicle left the roadway and struck a metal pole. The vehicle collision was observed by a security guard who rendered aid and was joined by a police officer and paramedics. The car fire was found to be arson.

| 4/29/1998 | **Raymond Nakovics, Firefighter** | **Career, Age 49** |
| | **New York City Fire Department, NY** | **Heart Attack** |

Firefighter Nakovics suffered a heart attack at the scene of a multiple-alarm highrise fire.

| 5/2/1998 | **Joseph Kroboth, Jr., Fire Police Captain** | **Volunteer, Age 59** |
| | **The Volunteer Fire Company of Halfway, MD** | **Struck by Vehicle** |

Captain Kroboth was directing traffic at the scene of a serious motor vehicle collision. The scene was very busy and a medical helicopter was in the process of landing. A pickup truck suddenly changed lanes and struck Captain Kroboth. The driver's attention was directed toward incident operations. According to the police report, Captain Kroboth was thrown 150 feet. He was wearing a reflective vest and utilizing a flashlight with safety wand. He died of massive head and chest injuries on 5/3/98.

| 5/2/1998 | **Michael A. Pizinger, Firefighter** | Career, Age 33 |
| | **Los Angeles City Fire Department, CA** | Drowning |

Firefighter Pizinger was killed when he drowned during a scuba diving training accident.

| 5/5/1998 | **Patrick Henry McKinney, Jr., Firefighter** | Volunteer, Age 72 |
| | **Colorado City Volunteer Fire Department, TX** | Apparatus Rollover |

Firefighter McKinney was driving a converted tanker from one brush fire to another when he lost control of the tanker as it crossed a narrow bridge. The tanker rolled three times after the right rear wheels of the vehicle caught on a concrete guardrail. Firefighter McKinney was ejected; another firefighter who was a passenger in the tanker was injured in the collision.

| 5/7/1998 | **Victor Clement Castillo, Fire Suppression Technician** | Career, Age 43 |
| | **El Paso Fire Department, TX** | Seizure Disorder |

Fire Suppression Technician Castillo was participating in a mandatory maze training exercise. During the event, Technician Castillo bumped his head twice, but told instructors that he was okay. After going off duty and going home, Technician Castillo became ill and was taken to the hospital by his wife. Later that month he was hospitalized for seizures and remained under a doctor's care until his death. He never returned to duty. The cause of death was ruled as cardiopulmonary arrest leading to anoxic brain injury. Underlying causes were aspiration secondary to a seizure, and seizure disorder secondary to his head injury. Further information related to this incident can be found in NIOSH Fire Fighter Fatality Investigation 99–F–08. Technician Castillo died on 1/21/1999.

| 5/7/1998 | **Jesus Mercado, Firefighter** | Career, Age 35 |
| | **Puerto Rico Fire Department, PR** | Apparatus Rollover |

Firefighter Mercado was killed in the rollover of a ladder truck while en route to a furniture warehouse fire in the town of Bayamon. One of the rear tires on the apparatus failed, resulting in the collision. Four other firefighters were injured.

| 5/9/1998 | **Daniel W. Mumford, Firefighter/Driver** | Volunteer, Age 48 |
| | **West Haverstraw Fire Department, NY** | Heart Attack |

Firefighter/Driver Mumford was the operator of a tower ladder apparatus at a mutual-aid structure fire. While preparing to leave the scene, Firefighter/Driver Mumford experienced an apparent heart attack. He had previously been under a doctor's care for a cardiac condition but had been released to drive a fire truck. Rain slicked roads and an earlier fall by Firefighter/Driver Mumford may have contributed to his death.

| 5/19/1998 | **Eugene Williard Blackmon, Jr., Firefighter** | Career, Age 38 |
| | **Chicago Fire Department, IL** | Drowning |

Firefighter Blackmon was conducting an underwater search for two reported drowning victims in the Calumet River. While going from shore to a boat he lost his grip on a flotation device and slipped under the water. He had removed his scuba tank prior to entering the water. Firefighter Blackmon was recovered after approximately 10–15 minutes and provided with emergency medical care. He was airlifted by a fire department helicopter to a local hospital but was pronounced dead at the hospital. Further information related to this incident can be found in NIOSH Fire Fighter Fatality Investigation 98–F–18.

| 5/26/1998 | **Jake M. Hoeffner, Junior Fire Department Captain** | Volunteer, Age 17 |
| | **Yaphank Fire Department, NY** | MVA |

Firefighter Hoeffner was a passenger in the bed of a fire department pickup as it proceeded across a parking lot during preparation for fire department training. Firefighter Hoeffner fell from the pickup and struck his head, sustaining fatal injuries. He was not wearing a seat belt. Firefighter Hoeffner died on 5/31/1998.

5/27/1998	**Walter A. Ernst, Firefighter**	**Volunteer, Age 61**
	East Meadow Fire Department, NY	**Heart Attack**

Firefighter Ernst assisted in the training of firefighters using SCBAs in a simulated smoke filled structure. He complained of pain in his left shoulder and was extremely fatigued at the conclusion of the training exercise as well as later in the fire station. He died at home in bed a few hours later of a heart attack.

5/29/1998	**Robert W. Munter, Fire Chief**	**Career, Age 56**
	Berlin Fire Department, MA	**Heart Attack**

Chief Munter, the only paid member of his department, was conducting an inspection of a new school that was under construction in his jurisdiction. At some point during the inspection, Chief Munter suffered a heart attack. He was found by construction workers lying prone with a head injury that likely occurred as he fell. Chief Munter was treated by members of his own department and transported to the hospital, where he was pronounced dead.

6/2/1998	**Dennis L. Buroker, Sergeant**	**Career, Age 44**
	Muncie Fire Department, IN	**Heart Attack**

Sergeant Buroker had gone to bed after responding to a brush fire at approximately 1:00 a.m. He was found dead in his bed by other firefighters in the morning. The cause of death was determined to be a heart attack due to hypertrophic cardiomyopathy (a genetic disease).

6/5/1998	**James Blackmore, Lieutenant**	**Career, Age 48**
	Scott J. LaPiedra, Captain	**Career, Age 40**
	New York City Fire Department, NY	**Trauma, Burns**

Along with other firefighters, Lieutenant Blackmore and Captain LaPiedra were conducting a search on the second floor of a commercial/residential structure. A civilian fire victim had been reported to be trapped in the area. Without warning, the second floor collapsed into the fire area on the first floor, trapping firefighters in a live fire on the first floor. Two firefighters died and four were seriously injured. The civilian fire victim had escaped through a back entrance. Lieutenant Blackmore was pronounced dead at the hospital after being recovered by other firefighters. The cause of death was crushing trauma and burns resulting in a heart attack. Captain LaPiedra suffered severe burns over 70 percent of his body and died on 7/4/1998. The cause of death was thermal burns resulting in cardiac arrest. More information related to this incident is available in NIOSH Fire Fighter Fatality Investigation 98–F–17.

6/26/1998	**Douglas L. Rohrbaugh, Fire Police Lieutenant**	**Volunteer, Age 52**
	Laurel Fire Company, PA	**Struck by Vehicle**

Fire Police Lieutenant Rohrbaugh was directing traffic at the scene of a motor vehicle collision. He was struck from the rear by a pickup truck. The pickup left the scene without stopping. Lieutenant Rohrbaugh was thrown 97 feet and landed along the side of the road.

6/27/1998	**Jerry David Donahue, Pilot**	**Wildland Contractor, Age 57**
	Charles Franklin Key, Copilot	**Wildland Contractor, Age 59**
	Neptune Aviation Services, MT;	**Aircraft Crash**
	Gila National Forest, NM	

Pilot Donahue and Copilot Key were killed as a result of the crash of their Lockheed SP–2H aircraft while fighting the Leggert wildland fire under contract for the United States Forest Service. The crash occurred about 5 miles west of Reserve, New Mexico. The tanker had completed a dry pass over the fire area, and then circled around to make a second pass and release its load. At that time, it contacted trees, crashed, and burned. The aircraft was carrying 2,450 gallons of fire retardant. More information related to this incident can be found in National Transportation Safety Board report FTW98GA86.

| 6/27/1998 | Johnnie Ray Park, Captain | Career, Age 43 |
| | Cullman Fire Department, AL | Heart Attack |

Captain Park died of a heart attack that occurred at the scene of a motor vehicle collision. The incident had been in progress for about an hour. Captain Park had assisted with patient treatment and scene cleanup. He was sitting in the cab of his engine and was beginning to complete an incident report when he was stricken. His crew immediately transported him to the hospital in the engine. The weather was hot and humid. Further information related to this incident can be found in NIOSH Fire Fighter Fatality Investigation 98–F–22.

| 6/29/1998 | Timothy D. Allen, Firefighter | Volunteer, Age 25 |
| | Central High Fire Department, OK | MVA |

Firefighter Allen was responding to his fire station in his personal vehicle when he was involved in a collision with another vehicle at an uncontrolled intersection. Firefighter Allen was thrown through the windshield of his vehicle and landed almost 20 feet away. Two children that were passengers in Firefighter Allen's vehicle were injured. The driver of the other vehicle was injured. No one in either car was wearing a seat belt.

| 7/6/1998 | Tulon Lee Goodwin, Firefighter/Forestry Worker | Wildland Career, Age 50 |
| | Alabama Forestry Commission, AL | Heart Attack |

Firefighter Goodwin was stricken with a heart attack at the scene of a wildland fire. Children playing with bottle rockets caused the fire. Firefighter Goodwin had been operating on the scene for about 4 hours and was plowing a fire line.

| 7/17/1998 | John Rochford Kennedy, Fire Police Officer | Volunteer, Age 75 |
| | Ocean Pines Volunteer Fire Department, MD | Heart Attack |

Fire Police Officer Kennedy was stricken with a heart attack as he directed traffic at the scene of a motor vehicle collision.

| 7/23/1998 | Matthew P. Casboni, Firefighter | Volunteer, Age 55 |
| | Saint John Volunteer Fire Department, IN | Heart Attack |

Firefighter Casboni died of a heart attack that occurred as he was acting as the air supply officer at a working structure fire.

| 7/23/1998 | Thomas E. Prendergast, Captain | Career, Age 56 |
| | Chicago Fire Department, IL | Heart Attack |

Captain Prendergast and his crew were fighting a two-alarm fire in a residential occupancy. Captain Prendergast and his crew were operating hose lines when he complained of chest pain and shortness of breath. He was immediately escorted to an ambulance and transported to the hospital. Captain Prendergast died as a result of a heart attack. He died on 8/8/1998.

| 8/1/1998 | Barvon Coy Hamilton, Firefighter | Volunteer, Age 71 |
| | Southern Oaks Volunteer Fire Department, TX | Hypovolemic Shock |

Firefighter Hamilton was on the back step of a pumper as it was relocated at the scene of a brush fire. The pumper was backing up, and Firefighter Hamilton was on the back step securing hose that had been reloaded. As the pumper backed up, Firefighter Hamilton attempted to warn the driver about a utility pole that was behind the apparatus. He apparently lost his grip and was crushed between the pumper and the pole. His right leg was amputated below the knee. Despite the efforts of local EMS providers, Firefighter Hamilton died of hypovolemic shock (loss of blood). Further information related to this incident can be found in NIOSH Fire Fighter Fatality Investigation 98–F–19.

| 8/3/1998 | Donald Claude Martin, Firefighter | Volunteer, Age 34 |
| | Van Buren Fire Department, ME | Heart Attack |

Firefighter Martin was in his fire station preparing to respond to assist with a search for a missing child. He became ill and was transported by other firefighters to the hospital. He was pronounced dead upon arrival. Firefighter Martin died of a heart attack.

| 8/16/1998 | Larry Joe King, Sr., Firefighter
Maury City Fire Department, TN | Paid–Call, Age 42
Heart Attack |

Firefighter King was attempting to pry open the hood of a pickup truck that was on fire. He suffered a heart attack and died.

| 8/18/1998 | Calvin Harbaugh, Sr., Fire Police Officer
Ebenezer Fire Company, PA | Volunteer, Age 56
Heart Attack |

Fire Police Officer Harbaugh was stricken with a heart attack while directing traffic at the scene of a motor vehicle collision.

| 8/20/1998 | John M. Walker, Private
Memphis Fire Department, TN | Career, Age 58
Heart Attack |

Private Walker was assigned to light duty to the department's SCBA maintenance shop. He was stricken with a heart attack while on duty. Despite immediate advanced life support care, Private Walker died.

| 8/29/1998 | Justin Melton, Firefighter
Scott Selby, Firefighter
Marks Fire Department, MS | Volunteer, Age 22
Volunteer, Age 35
Smoke Inhalation |

Firefighters Melton and Selby were working in different areas of a structure fire that involved a commercial building. A collapse occurred which trapped Firefighter Melton as he and other firefighters were advancing a hoseline on the fire. Firefighter Selby was on the roof of the fire structure attempting ventilation when he fell into the fire area and was killed. Both firefighters died of asphyxiation due to smoke inhalation. Further information related to this incident can be found in NIOSH Fire Fighter Fatality Investigation 98–F–21.

| 8/29/1998 | Robert F. Peters, Lieutenant
Hastings on Hudson Fire Department, NY | Volunteer, Age 52
Heart Attack |

Lieutenant Peters was completing paperwork after returning from a response to an automatic fire alarm. Lieutenant Peters had driven an aerial apparatus to the incident. He was stricken with a heart attack and died.

| 8/31/1998 | Brian Carrasco, Inmate Handcrew Firefighter
Los Angeles County Fire Department, CA | Wildland Part-Time, Age 35
Apparatus Rollover |

Firefighter Carrasco was killed in a vehicle collision while working as a member of an inmate handcrew. The brush/engine vehicle in which we was riding rolled over. Eleven others were injured.

| 9/4/1998 | Juan Manuel Hernandez Jr., Firefighter
United States Department of Agriculture Forest
Service, NM | Wildland Full-Time Seasonal,
Age 24
Apparatus Rollover |

Firefighter Hernandez was killed in a vehicle collision while working near Willows, California. The engine in which he was riding was stuck by a pickup truck that crossed the centerline and impacted the engine along the left underside and rear dual wheels. The engine flipped and landed upside down. Firefighter Hernandez was wearing a seat belt but was partially ejected from the vehicle. He was pronounced dead at the scene.

| 9/4/1998 | Allen Lawrence Heirtzler, Firefighter
Slaughter Volunteer Fire Department, LA | Volunteer, Age 22
MVA |

Firefighter Heirtzler was responding to a trailer fire in his role as a volunteer firefighter. He was responding in his Sheriff's Department cruiser when he was involved in a collision. Firefighter Heirtzler's vehicle was traveling at a high rate of speed when it skidded, struck a cow in the roadway, crashed into a utility pole, rolled over, and burned. Firefighter Heirtzler was not wearing a seat belt at the time of the collision.

9/5/1998   **Eugene P. McDonough, Firefighter**   **Career, Age 54**
                  Saint Johnsbury Fire Department, VT   **Trauma**

Firefighter McDonough responded with other members of his department to a mutual-aid fire in a recycling facility. While opening a large door to allow a master stream attack, Firefighter McDonough was crushed when a parapet wall collapsed. The cause of the fire was arson. Further information related to this incident can be found in NIOSH Fire Fighter Fatality Investigation 98–F–20.

9/9/1998   **Ernest Alan McElroy, Forest Ranger II**   **Wildland Career, Age 40**
                  Arkansas Forestry Commission, AR   **Overrun by Wildfire**

Forest Ranger McElroy was plowing a fire line with a bulldozer. The fire overcame his position. He attempted to back out but struck a tree. Ranger McElroy dismounted the bulldozer and proceeded down the fire line on foot. He was burned over 60 percent of his body but still managed to walk the 1/2 mile to a waiting ambulance. Ranger McElroy died of his injuries on 10/28/1998. Further information related to this incident can be found in NIOSH Fire Fighter Fatality Investigation 98–F–30.

9/14/1998   **Randy Sims, Captain**   **Volunteer, Age 44**
                  Antioch Volunteer Fire Department, SC   **Heart Attack**

Captain Sims was at the scene of a structure fire assisting with overhaul when he suffered a heart attack and died.

9/18/1998   **Donald Trotochaud, Firefighter/Senior Airman**   **Career, Age 23**
                  Laughlin Air Force Base, TX   **MVA**

Firefighter Trotochaud was killed in a single vehicle collision while en route to standby at a remote airfield. One other firefighter was injured.

9/21/1998   **David M. Brinkley, Firefighter/Past Chief**   **Volunteer, Age 43**
                  United Communities Volunteer Fire Department, MD   **Heart Attack**

Firefighter Brinkley suffered a heart attack while refilling SCBA cylinders after a response to a vehicle fire.

9/24/1998   **Tony B. Chapin, Firefighter/EMT**   **Volunteer, Age 19**
                  Willamina Fire Department, OR   **MVA**

Firefighter Chapin was killed while on the way to a paramedic training class in his personal vehicle. A car crossed the centerline and struck the vehicle that Firefighter Chapin was driving. He survived the initial impact but died the next day. Firefighter Chapin was wearing his seat belt.

9/27/1998   **Preston Edgar Patterson, Firefighter/Fire Police**   **Volunteer, Age 66**
                  The Manchester Fire Engine and Hook and Ladder   **Heart Attack**
                  Company Number One, MD

Firefighter Patterson was stricken with a heart attack while performing fire police duties at the scene of a motor vehicle collision.

9/28/1998   **Neil A. Holmes, Captain**   **Career, Age 55**
                  Fresno City Fire Department, CA   **Brain Aneurysm**

Captain Holmes was found unconscious in the restroom of his fire station. He suffered a brain aneurysm.

9/28/1998   **Paul P. Satterfield, Battalion Chief**   **Career, Age 60**
                  Nashville Fire Department, TN   **Brain Aneurysm**

While in command of a fire, Chief Satterfield complained of illness. He finished his shift and went off duty in the morning. He died at home on 9/29/98 of a brain aneurysm.

9/30/1998   **Robert Odell Lee, Fire Chief**   **Volunteer, Age 56**
                  North River Valley Volunteer Fire Company, WV   **Heart Attack**

Chief Lee was monitoring pump operations during a drill at a nursing home. He was wearing full protective clothing but no SCBA. He was not performing any strenuous activity; however, the day was hot and humid. Chief Lee suddenly collapsed and died of a heart attack. Further information related to this incident can be found in NIOSH Fire Fighter Fatality Investigation 98–F–11.

10/5/1998	**Gary D. Nagel, Airtanker Pilot**	**Wildland Contractor, Age 62**
	San Joaquin Helicopters, CA	**Aircraft Crash**

Airtanker Pilot Nagel was killed in the crash of a Grumman TS–2A airtanker when he misjudged his maneuvering altitude and impacted the terrain. He had made two drops on the Mount Edna fire near Banning, California, and was preparing to make a third. Other factors that contributed to the crash were the mountainous terrain, tailwind conditions, and turbulence in the area. Airtanker Pilot Nagel was an employee of San Joaquin Helicopters, a contractor to the California Department of Forestry and Fire Protection. More information related to this incident can be found in National Transportation Safety Board report LAX99GA005.

10/5/1998	**Thomas Oscar Wall, Captain**	**Career, Age 44**
	Orange County Fire Authority, CA	**Heart Attack**

Captain Wall died at the Taylor fire in Riverside, California. He was protecting exposed dwellings when he told other firefighters that he did not feel well and collapsed. Captain Wall died of a heart attack.

10/13/1998	**Barry L. Wary, Firefighter**	**Volunteer, Age 51**
	Klingerstown Volunteer Fire Company, PA	**Heart Attack**

Firefighter Wary was actively involved in the suppression of a fire in an industrial occupancy. Upon exiting the structure, he collapsed and died of a heart attack.

10/24/1998	**Carson L. Gosey, Sr., Firefighter/Fire Police**	**Volunteer, Age 60**
	Shiloh/Danieltown/Oakland Fire Department, NC	**Struck by Vehicle**

A vehicle at the scene of a training exercise struck Firefighter Gosey as he assisted a water tanker that was crossing the road.

10/24/1998	**Lawrence D. Thrower, Lieutenant**	**Volunteer, Age 51**
	Sidney Fire Department, NY	**Heart Attack**

Lieutenant Thrower responded to the scene of a dumpster fire at a manufacturing facility. He was equipped in full protective clothing and SCBA as he and his crew extinguished the fire and performed overhaul. Lieutenant Thrower removed his facepiece at the conclusion of operations and was beginning to remove his other protective clothing when he collapsed. Firefighters initiated advanced life support care and Lieutenant Thrower was transported to the hospital. He was pronounced dead of a heart attack a short time later.

11/6/1998	**Hubert Sidney Jones, Fire Chief**	**Volunteer, Age 29**
	Thoroughfare Volunteer Fire Department, NC	**Carbon Monoxide/ Smoke Inhalation**
	Robby Dean Blizzard, First Lieutenant	**Volunteer, Age 24**
	Arrington County Volunteer Fire Department, NC	**Smoke Inhalation**

Chief Jones and First Lieutenant Blizzard were killed as they fought a fire in an automobile salvage yard storage building. Firefighters believed that they had found the seat of the fire and were applying water when a rapid change in conditions occurred. Chief Jones ran out of air while trying to escape. Lieutenant Blizzard entered the structure to search for Chief Jones. He ran out of air, became disoriented, and failed to exit the building. Lieutenant Blizzard was wearing a PASS device but it was not activated. Chief Jones was not equipped with a PASS device. The causes of death for Chief Jones were listed as carbon monoxide poisoning and smoke inhalation and the cause of death for Lieutenant Blizzard was listed as carbon monoxide poisoning. Lieutenant Blizzard was also a career firefighter in another community but was off duty at the time. Further information related to this incident can be found in NIOSH Fire Fighter Fatality Investigation 98–F–32.

11/7/1998   Paul Allen Laux, Captain   Volunteer, Age 50
                Cool Spring Township Fire Department, IN   Cardiac

Captain Laux drove a water tanker to the scene of a reported structure fire. When the report turned out to be steam, the incident commander released all units to return to station. As reports were completed and units prepared to return to station, a firefighter noticed that the door to the tanker was open. As he looked inside, he observed Captain Laux slumped over the wheel and unconscious. Captain Laux was removed from the tanker and provided with basic life support care until the arrival of an advanced life support ambulance. No defibrillator was available initially but one was utilized upon the arrival of ALS. Captain Laux had a history of cardiovascular disease including bypass surgery but had been cleared for fire fighting by his personal physician. Further information related to this incident can be found in NIOSH Fire Fighter Fatality Investigation 99–F–05.

11/8/1998   Charles Peter Frank III, Deputy Chief   Volunteer, Age 56
                West Weatherfield Fire Department, VT   Heart Attack

Deputy Chief Frank was in command of a vehicle fire on a local highway. He participated in forcible entry, water supply, and fire attack. After the fire was extinguished, Chief Frank began to speak with the people who reported the fire when he suddenly collapsed. Despite immediate EMS assistance on the scene and the arrival of advanced life support care, Chief Frank died of a heart attack.

11/9/1998   Thomas Benjamin Rice, Fire Police Officer   Volunteer, Age 70
                Village of Perry Fire Department, NY   Heart Attack

Fire Police Officer Rice had been directing traffic at the scene of a commercial structure fire. His duties were completed and he had been released. A passing firefighter noticed Officer Rice on the ground near his vehicle. Despite immediate aid, he died of a heart attack. Nothing unusual preceded the attack. Officer Rice had a history of heart problems.

11/13/1998 William Dwight Yankey, Firefighter   Career, Age 35
                Harrodsburg Fire Department, KY   Pulmonary Embolism

Firefighter Yankey's captain was awakened when Firefighter Yankey fell out of bed. His captain found him not breathing and began CPR with the assistance of other firefighters. Firefighter Yankey regained consciousness at least twice and spoke with other firefighters. He was transported to the hospital and was pronounced dead. The cause of death was listed as acute cardio-respiratory failure due to occlusive pulmonary thromboemboli (blood clot in the lungs).

11/18/1998 Roger DeWayne Bookout, Heavy Fire Equipment   Wildland Career, Age Unknown
                  Operator   Apparatus Rollover
                California Department of Forestry and Fire
                  Protection, CA

Operator Bookout was killed in an unwitnessed rollover of the tractor that he was using to perform fire lookout road maintenance.

11/24/1998 Norman Neal Almond, Captain   Career, Age 46
                Craig Daniel Brown, Driver/Operator   Career, Age 27
                Parsons Fire Department, KS   Electrocution

A painting contractor found that their ladders were not long enough to reach the upper portion of the church. A representative of the church contacted city hall and requested an aerial ladder. After an assessment by the fire chief, a reserve pumper was sent to the scene to allow for the use of its ground ladders. Captain Almond and Driver/Operator Brown were assisting with the positioning of a ladder on the exterior of a church that was being painted. The aluminum ladder made contact with an electrical service line, resulting in the fatal electrocution of both firefighters and injury to one additional firefighter. The firefighters were positioning the ladder since it was too cumbersome for the two painters to position by themselves. Further information related to this incident can be found in NIOSH Fire Fighter Fatality Investigation 98–F–31.

| 12/2/1998 | Steven C. Mayfield, Firefighter | Career, Age 47 |
| | Houston Fire Department, TX | Heart Attack |

Firefighter Mayfield was participating in Federal Aviation Administration mandated Aircraft Rescue Fire Fighting training at the Dallas–Fort Worth airport training facility. Firefighter Mayfield was in the interior of an aircraft fuselage mockup lifting and pulling a mannequin when he experienced a fatal heart attack.

| 12/9/1998 | Steve Austin Tippins, Assistant Chief | Volunteer, Age 37 |
| | Etolie Volunteer Fire Department, TN | MVA |

Chief Tippins was responding in his personal vehicle to an EMS incident. He failed to yield at a stop sign and was broadsided on the driver's side by a truck. Chief Tippins was killed instantly. His wife, a passenger in his vehicle, was injured.

| 12/12/1998 | Stephen E. Gessler, First Assistant Chief | Volunteer, Age 44 |
| | Little Falls Fire Department, NJ | Cardiac |

Chief Gessler was in command of a search effort to find a missing civilian. The man was located and Chief Gessler ordered an ambulance to transport him. Immediately after giving the order, Chief Gessler complained of being dizzy and collapsed. Bystanders initiated resuscitation efforts immediately. Chief Gessler died of dilated cardiomyopathy (a disorder in which the heart muscle is weakened and cannot pump blood efficiently).

12/18/1998	James F. Bohan, Firefighter	Career, Age 25
	Christopher M. Bopp, Firefighter	Career, Age 27
	Joseph P. Cavalieri, Lieutenant	Career, Age 42
	New York City Fire Department, NY	Trapped

Firefighter Bohan, Firefighter Bopp, and Lieutenant Cavalieri were killed while fighting a residential highrise structure fire. As they rushed to the tenth floor to search for victims, they were overcome by a wave of heat and smoke that killed all three. The heat wave, or fireball, may have been propelled by a gust of wind coming through the fire apartment. The automatic closing device on the apartment door had been removed or had malfunctioned. The building's hallway sprinklers did not activate due to a closed valve. Six other firefighters were injured in the fire.

| 12/18/1998 | Thomas J. Concannon, Fire Police Lieutenant | Volunteer, Age 55 |
| | Wormleysburg Fire Company #1, PA | Heart Attack |

Fire Police Lieutenant Concannon responded to a motor vehicle collision. He parked his vehicle short of the incident scene. Members working on the incident scene assumed that he had stopped in that location to divert traffic. At the conclusion of the incident, Lieutenant Concannon failed to respond to a radio call. Shortly thereafter, he was found slumped over the steering wheel of his vehicle, unresponsive, with the vehicle still in gear and his foot on the brake. Despite immediate medical care, Lieutenant Concannon died of a heart attack. Further information related to this incident can be found in NIOSH Fire Fighter Fatality Investigation 99-F-10.

| 12/31/1998 | Kennon Loy Williams, Captain | Volunteer, Age 27 |
| | Banks County Fire Department, GA | Trapped |

Captain Williams and other members of his department were conducting an offensive attack on an arson fire of a church built circa 1850. Captain Williams was caught under heavy timbers in a roof collapse. Further information related to this incident can be found in NIOSH Fire Fighter Fatality Investigation 99-F-04.

| 1/5/1999 | Carl Arnold Olsen, Firefighter | Volunteer, Age 51 |
| | Kiln Volunteer Fire Department, MS | Struck by Vehicle |

Firefighter Olsen was killed when a car struck him as he worked to refill a water tanker. The lights of the tanker blinded the driver of the car. Firefighter Olsen was not wearing any reflective material at the time he was struck. Firefighter Olsen's wife is also a volunteer firefighter and was on the scene at the time of the fatal incident.

1/8/1999   **Daniel J. O'Connell, Firefighter**   **Volunteer, Age 58**
          Putnam Lake Fire Department, NY   **Heart Attack**

Firefighter O'Connell responded with his department to a report of a structure fire. The fire involved an oil burner in a furnace and was extinguished with the use of a dry chemical extinguisher. Firefighter O'Connell, who had some experience with this type of equipment, was ordered to investigate further and make sure that the fire was out. He was exposed to soot and dry chemical residue. He left the house and collapsed on the front lawn in cardiac arrest. Firefighters initiated EMS treatment immediately, and Firefighter O'Connell was transported to the hospital. It is the opinion of Firefighter O'Connell's doctor that the heart attack was caused by exposure to smoke and chemicals. Firefighter O'Connell suffered from injuries to the brain caused by a lack of oxygen and died on 1/25/1999. No autopsy was performed.

1/9/1999   **Jason A. Gouckenour, Firefighter**   **Volunteer, Age 22**
          Worthington–Jefferson Volunteer Fire Department, IN   **Trapped**

Firefighter Gouckenour entered a structural fire in a house alone with a hose line. He was equipped with full turnout gear and SCBA, but was not equipped with a PASS device. It is believed that he tripped over a coffee table and became entangled in a couch. He removed his SCBA to call for help and was overcome by extremely heavy heat and smoke conditions. Firefighters on the scene attempted a rescue but were driven back by intense heat and flames and finally by the collapse of the house's roof. Firefighter Gouckenour's body was found approximately 10 feet inside the front door of the structure. The cause of death was asphyxiation due to smoke inhalation and carbon monoxide. Firefighter Gouckenour joined the fire department after his home had burned 2 years before his death. Additional information about this incident can be found in NIOSH Fire Fighter Fatality Investigation 99–F–02.

1/10/1999   **Tracy Dolan Toomey, Firefighter**   **Career, Age 52**
          Oakland Fire Department, CA   **Trauma**

Firefighter Toomey was crushed and killed when the second floor of a turn-of-the-century residential structure collapsed into the first floor. The fire eventually went to six alarms. A total of four firefighters were trapped by the collapse, including Firefighter Toomey. Additional information about this incident can be found in NIOSH Fire Fighter Fatality Investigation 99–F–03.

1/11/1999   **Martin Richard Wauson, Forestry Technician**   **Wildland Part-Time, Age 55**
          United States Department of Agriculture Forest   **Cardiac**
          Service, AR

Forestry Technician Wauson was participating in an annual requalification test for seasonal firefighting duty with the U.S. Forest Service (USFS). The test required the participant to carry a 45-pound backpack and cover 3 miles within 45 minutes. Forestry Technician Wauson had just passed the 1-mile mark when he collapsed. Medical personnel at the site immediately provided CPR and defibrillation; however, he was not revived. The autopsy revealed that Forestry Technician Wauson's death was caused by hypertensive arteriosclerotic disease. As a result of his death, the USFS temporarily suspended use of the pack test.

1/16/1999   **James R. Tolan, Firefighter**   **Career, Age 55**
          Dunmore Fire Department, PA   **Heart Attack**

Firefighter Tolan had responded to a report of a structural fire as the driver and sole occupant of a ladder truck. When he arrived on the scene, he exited his apparatus and told his shift commander that he was not feeling well. He was taken to the hospital by ambulance, was treated for a heart attack, and subsequently was discharged from the hospital. Firefighter Tolan never returned to work. He became ill again in June, was readmitted to the hospital, and died on 6/9/1999.

1/19/1999	**James H. McGroarty, Firefighter/Fire Investigator**	**Career, Age 43**
	Syracuse Fire Department, NY	**Trauma**

Fire Investigator McGroarty was in the attic of a residential structure that had experienced a fire 5 days before. A private fire investigator and an electrical consultant also were in the attic with Investigator McGroarty. During the course of the investigation, a chimney, which had been supported by the roof prior before the fire, collapsed onto Investigator McGroarty causing severe injuries. The chimney was too heavy for the personnel on the scene to lift and it stayed in place until additional personnel arrived. Medical treatment was initiated immediately after Investigator McGroarty was freed. He was transported to the hospital where he died. The cause of death on the autopsy was listed as multiple injuries caused by falling debris from a recent fire. Additional information about this incident can be found in NIOSH Fire Fighter Fatality Investigation 99–F–06.

1/27/1999	**Ralph J. Loyd, Firefighter**	**Volunteer, Age 90**
	Greenville Fire Protection District, IL	**Heart Attack**

Firefighter Loyd was a member of the Greenville Fire Protection District and its predecessors for 70 years. He drove a fire department equipment truck to the scene of a reported grain elevator fire. As he was standing by the truck shortly after arrival, he collapsed from a heart attack. Medical aid was rendered at the scene. Firefighter Loyd died the next day; no autopsy was performed.

1/29/1999	**Joseph R. "Dick" Murphy, Firefighter**	**Career, Age 64**
	Boston Fire Department, MA	**Heart Attack**

Firefighter Murphy was preparing to back his command vehicle into the station after returning from a report of smoke in an apartment building. Firefighter Murphy did not complain of any illness at the scene. As he prepared to back in, he collapsed on the steering wheel from a heart attack. The chief officer who was riding in the vehicle attempted to shift the vehicle into park but was unsuccessful. The command vehicle, a Chevrolet Suburban, proceeded in reverse and collided with a pumper located in the apparatus bay. Despite the immediate administration of CPR and defibrillation by firefighters and the subsequent arrival of advanced life support personnel, Firefighter Murphy was pronounced dead shortly after arriving at the hospital. The autopsy listed the cause of death as occlusive coronary heart disease. Additional information about this incident is contained in NIOSH Fire Fighter Fatality Investigation 99–F–12.

2/9/1999	**Gerald "Jerry" Myers, Fire Safety Officer**	**Volunteer, Age 59**
	Sumpter Fire Department, OR	**Struck by Apparatus**

Firefighter Myers was clearing snow from around fire hydrants with a backhoe. A mechanical problem occurred with the backhoe and Firefighter Myers brought the equipment to a repair facility to fix the problem. During the course of the repairs, Firefighter Myers was struck by a part of the equipment and died of a traumatic injury.

2/12/1999	**Jimmy "Wayne" Kittle, Captain**	**Career, Age 49**
	Whitfield County Fire Department, GA	**Pulmonary Embolism**

Captain Kittle suffered a massive pulmonary embolism while on duty at the fire station and died.

2/15/99	**Brian William Collins, Assistant Fire Chief**	**Volunteer, Age 35**
	Phillip Wayne Snell–Dean, Captain	**Volunteer, Age 29**
	River Oaks Volunteer Fire Department, TX
	Gary Charles Sanders, Firefighter	**Volunteer, Age 20**
	Sansom Park Fire Department, TX	**Smoke Inhalation/Burns**

Chief Collins, Captain Dean, and Firefighter Sanders were members of an attack team working in the interior of a church that was involved in fire when the roof collapsed and trapped the three in the fire area. The fire was set in a shed next to the church and spread into the attic rafters of the church building itself. Firefighters were attacking an attic fire from the interior of the structure and were being ordered to the exterior as the collapse occurred. Firefighters were on the roof at the time of the collapse, and one fell into the church but was not seriously injured. The fire was determined to be caused by arson. Chief Collins died of extensive thermal burns. He had a carboxyhemoglobin level of 6.25 percent. Captain Dean died of smoke inhalation with thermal injuries. He had a carboxyhemoglobin level of 52.5 percent. Firefighter Sanders died of smoke inhalation with thermal injuries. He had a carboxyhemoglobin level of 72.5 percent. Chief Collins and Captain Dean were a career officers with the Fort Worth Fire Department.

2/15/1999	**Terry Lee Myers, Driver/Operator**	**Volunteer, Age 50**
	Vigilant Hose Company, Emmitsburg, MD	**Heart Attack**

Driver/Operator Myers was working as a pump operator at the scene of a brush fire on the campus of Mount Saint Mary's College in Emmitsburg, Maryland. He had been working for about 45 minutes when he collapsed of a heart attack. Driver/Operator Myers had not complained of any sickness prior to his attack. Members of his department and paramedics from the rescue squad provided emergency care. Despite their efforts, he was pronounced dead at a local hospital. The brush fire was caused by the spread of an unattended fire, which had been set to dispose of cleared brush and trees. Additional information about this incident can be found in NIOSH Fire Fighter Fatality Investigation 99–F–43.

2/16/1999	**Robert C. Stanmire, Sr., Firefighter**	**Volunteer, Age 52**
	Forest Grove Volunteer Fire Company, NJ	**Heart Attack**

Firefighter Stanmire suffered a heart attack as he and other members of his department prepared to respond to a suspicious brush fire. He was stricken in the fire station as he boarded a piece of fire apparatus while dressed in full turnouts. Fellow firefighters began CPR immediately, and the local rescue squad provided advanced life support care, but Firefighter Stanmire was dead upon arrival at a local hospital; no autopsy was performed.

2/18/1999	**Burton Frank Chestnut, Firefighter**	**Volunteer, Age 67**
	Brazos Volunteer Fire Department, TX	**Heart Attack**

Firefighter Chestnut was on a personal errand with two other Brazos firefighters when they came upon a fire in a woodpile that was against a structure. Firefighter Chestnut remained at the scene to begin notifying residents of the fire while other firefighters went to retrieve their fire apparatus. When the other firefighters returned, they found Firefighter Chestnut dead of an apparent heart attack.

2/19/1999	**Terry "Ted" Oliver, Assistant Fire Chief**	**Volunteer, Age 58**
	Eaton Rapids Fire Department, MI	**Heart Attack**

Assistant Chief Oliver was directing the overhaul and salvage of a bedroom fire that had been extinguished. He ascended the stairs to the second floor of the house to check on the progress of work in the fire area. Upon reaching the top of the stairwell, he collapsed and died of an apparent heart attack. He was removed from the structure and taken to the hospital by an ambulance that was on the scene of the fire. Assistant Chief Oliver was the first firefighter to die in the line of duty for the 125-year-old department.

2/28/1999	**Arthur Bruce Franklin, Firefighter**	**Volunteer, Age 50**
	Anderson County Fire Protection District, KY	**Unknown**

Firefighter Franklin died after returning home from a fire.

2/28/1999	**Alan W. Ducheck, Captain**	**Paid-on-Call, Age 46**
	DeSoto Fire and Rescue, MO	**Heart Attack**

Captain Ducheck was assisting with a lengthy vehicle extrication when he suffered a heart attack. He died on 3/1/1999.

3/12/1999	**Jerome Taylor, Captain**	**Volunteer, Age 69**
	Hillburn Fire Department, NY	**Heart Attack**

Captain Taylor collapsed and died of an apparent heart attack while directing traffic at a structure fire.

3/16/1999   **David L. Packard, Firefighter**   **Career, Age 56**
                Boston Fire Department, MS   **Heart Attack**

Firefighter Packard was preparing for duty when his engine company was dispatched to a motor vehicle collision. Firefighter Packard did not respond on the call because he was on the oncoming shift and the call was handled by the off-going shift. He was found by other firefighters upon their return to quarters. He had collapsed in the bunkroom and suffered an apparent heart attack. Despite immediate aid by on-scene firefighters, the use of a defibrillator, and advanced life support treatment by paramedics, Firefighter Packard died. No autopsy was performed. The cause of death as listed on his death certificate was asystole caused by coronary artery disease. Additional information about this incident can be found in NIOSH Fire Fighter Fatality Investigation 99–F–17.

3/16/1999   **Charles James "Chuck" Vodak, Firefighter**   **Volunteer, Age 45**
                Dunning Fire Department, NE   **Heart Attack**

Firefighter Vodak and other members of his department had been fighting a prairie fire for more than 4 hours. Firefighter Vodak complained of chest pains and experienced a heart attack. Other firefighters performed CPR on Firefighter Vodak for 1 hour as they waited for an ambulance to arrive at the isolated fire location. The path of the fire was 10 miles wide in some places and it eventually consumed 70,000 acres.

3/17/1999   **Walter J. Flyntz, Firefighter**   **Career, Age 44**
                Bridgeport Fire Department, CT   **Cardiac**

Firefighter Flyntz responded as part of an engine company to a report of a fire in the basement of an apartment building that was undergoing renovation. Upon arrival, he assisted with the connection of his engine to a hydrant and then helped check for fire extension on the upper floors of the building. After clearing the second floor, Firefighter Flyntz went to the third floor to check for extension. Shortly after his arrival on the third floor, he collapsed. A renovation worker discovered him and notified other firefighters. The cause of death was listed as atherosclerotic cardiovascular disease. Additional information about this incident can be found in NIOSH Fire Fighter Fatality Investigation 99–F–18.

3/23/1999   **Paul Haislopp, Captain**   **Volunteer, Age 50**
                Centerville Fire Department, OH   **Heart Attack**

Captain Haislopp experienced a heart attack at his fire station as he moved fire apparatus in preparation for a response to a structure fire.

4/2/1999   **Aubrey R. Tillman, Firefighter**   **Career, Age 57**
                Charleston Fire Department, SC   **Heart Attack**

Firefighter Tillman had been on duty for 14 hours and had participated in physically demanding training for 3-1/2 hours of his shift. At approximately 10 p.m., Firefighter Tillman experienced a temporary loss of consciousness. As other firefighters came to his aid, Firefighter Tillman regained consciousness and complained of severe chest pains. He was transported by ambulance to the hospital under paramedic care. His condition deteriorated in the ambulance and further deteriorated in the hospital emergency room. Firefighter Tillman died at 11:37 p.m. The death certificate listed "probable acute myocardial infarction" as the immediate cause of death. No autopsy was performed. Additional information about this incident can be found in NIOSH Fire Fighter Fatality Investigation 99–F–15.

4/6/1999   **Kenneth Allen Nickell, Captain**   **Volunteer, Age 28**
                Kevin Rex Smith, Firefighter/EMT   **Volunteer, Age 30**
                Route 377 Volunteer Fire Department, KY   **Overrun by Wildfire**

Captain Nickell and Firefighter Smith responded to a wildland fire in the Daniel Boone National Forest near Cranston, Kentucky. They were part of a seven-person team that was constructing a fire line in hardwood leaf litter on the forest floor. Nickell and Smith were in the lead and using a rake and a gasoline-powered leaf blower to construct the line. As the fire line was being constructed, spot fires were breaking over the fire line and several members of the team doubled back to control the spot fires. Captain Nickell and Firefighter Smith continued to construct the fire line. The fire was growing in intensity and the wind was picking up so the crew leader gave the order for all firefighters to pull back. Captain Nickell acknowledged the order and indicated that he and Firefighter Smith would pull back. Shortly thereafter, another radio transmission was received from Captain Nickell indicating that he and Firefighter Smith were burned or on fire. Evidence suggests that the two tried to outrun the fire uphill but were slowed by terrain. It appeared as if the firefighters attempted to run back through the fire to reach the burned area. At some point, they succumbed to the flames and collapsed. The cause of death for both firefighters was listed as asphyxia due to environmental oxygen deprivation, smoke inhalation, and acute carbon monoxide poisoning. Neither firefighter was equipped with a fire shelter. Additional information about this incident can be found in NIOSH Fire Fighter Fatality Investigation 99–F–14. A report also is available from the Kentucky Division of Forestry entitled *Report of the Accident Investigations Team for the Island Fork Fire, April 6, 1999, Near Cranston, Kentucky.*

| 4/8/1999 | John E. Murphy, Deputy Fire Chief | Paid-on-Call, Age 64 |
| | Russell Fire Department, MA | Cardiac |

Deputy Fire Chief Murphy was engaged in fighting a wildland fire for 4 hours. He collapsed and was down for 10 to 15 minutes before being discovered by another firefighter. The firefighter initiated CPR and transported him to a waiting ambulance in the bed of the fire chief's pickup (due to terrain). CPR continued through transport in the BLS ambulance, ALS treatment, and treatment in the emergency room. Despite all efforts, Chief Murphy died in the emergency room. The cause of death was listed as coronary atherosclerosis. Additional information about this incident can be found in NIOSH Fire Fighter Fatality Investigation 99–F–32.

| 4/12/1999 | Phillip M. Pinkowski, Jr., Firefighter | Volunteer, Age 59 |
| | Clarendon Volunteer Fire Department, VT | Heart Attack |

Firefighter Pinkowski collapsed and died of an apparent heart attack while acting as a pump operator at a residential structure fire.

| 4/15/1999 | Robert D. Peters, Firefighter | Volunteer, Age 71 |
| | West Lake Fire Department, PA | Stomach Aneurysm |

Firefighter Peters responded to an ambulance call and was not feeling well at the call. After the call, he went home. That afternoon, a relative found him at home nearly unconscious. Despite surgery, Firefighter Peters died as the result of a stomach aneurysm.

| 4/28/1999 | David J. Watts, Captain | Paid-on-Call, Age 56 |
| | Nantucket Fire Department, MA | Cardiac |

Captain Watts was engaged in active structural firefighting for 2-1/2 hours in a multiple-occupancy wood-frame building that was constructed circa 1849. After the fire was controlled, Captain Watts returned home and suffered a heart attack in the shower. CPR was administered, first by his wife and later by firefighters, and an automatic defibrillator was used by other firefighters who responded to Captain Watts' residence. He was transported by ambulance to a local hospital where he was stabilized. Captain Watts was transferred to a hospital in Boston by air but he never regained consciousness. He died on 5/2/1999; the cause of death was listed as coronary thrombosis and ventricular tachycardia. Additional information about this incident can be found in NIOSH Fire Fighter Fatality Investigation 99–F–19.

| 5/2/1999 | Kenneth Alan Strain, Firefighter | Volunteer, Age 28 |
| | Hemby Bridge Volunteer Fire Department, NC | Trauma |

Firefighter Strain was the sole occupant and driver of a 1996 pumper. He was returning to the station after a cancelled call for a motor vehicle collision. He pulled the pumper to the right side of the road to allow for the passage of other traffic when the right rear wheels of the pumper left the paved surface of the road and fell into a ditch. Firefighter Strain was unable to regain control, and the pumper struck a tree. Firefighter Strain was killed instantly and was pronounced dead at the scene. According to the highway patrol, speed was not a factor. Firefighter Strain was wearing his seat belt. It took other firefighters 2 hours to free him from the wreckage. Additional information about this incident can be found in NIOSH Fire Fighter Fatality Investigation 99–F–16.

5/3/1999   **Eric Noel Casiano, Firefighter**   **Career, Age 41**
                Philadelphia Fire Department, PA   **Trauma**

Firefighter Casiano and his company were fighting a structural fire in a residential occupancy. During the fire, Firefighter Casiano fell through a floor but appeared to be uninjured. After his company had been released and was back at quarters, everyone returned to bed. A short time later, Firefighter Casiano's company was dispatched to another call. Other firefighters found him down and without vital signs. Despite immediate CPR and ALS care within 4 minutes, Firefighter Casiano died. His autopsy revealed that he had a massive hemorrhage within his back muscles, which damaged his spinal cord. The hemorrhage was caused by the fall.

5/4/1999   **Arthur A. Tullis, Fire Chief**   **Part-Time (Paid), Age 61**
                LaGrange Park Volunteer Fire Department, IL   **Heart Attack**

Chief Tullis and members of his fire department and a neighboring department had responded to an automatic fire alarm in a retirement home. Chief Tullis was first on the scene and was exiting the building to command the arrival of other units when he collapsed from an apparent heart attack. Firefighters provided advanced life support care. The cause of death was heart disease. Carpet installers caused the alarm activation.

5/14/1999   **Lewis Edward Williams, Fire and Rescue Captain**   **Volunteer, Age 47**
                Fort Oglethorpe Fire & Rescue, GA   **Heart Attack**

Captain Williams worked for 2-1/2 hours on the scene of a trench collapse that trapped one worker. He participated in various tasks, including the unloading of supplies, and command. He collapsed at the command post without warning or any complaint of sickness. Despite immediate advanced life support care, Captain Williams died. The cause of death was listed as a heart attack. The trapped construction worker was rescued successfully. Additional information about this incident can be found in NIOSH Fire Fighter Fatality Investigation 99–F–49.

5/30/1999   **Lewis Jefferson Matthews, Firefighter**   **Career, Age 29**
                Anthony Sean Phillips, Sr., Firefighter   **Career, Age 30**
                District of Columbia Fire Department, Washington DC **Burns**

Firefighters Matthews and Phillips were members of two different engine companies working on the first floor of a townhouse fire. Both crews had entered the front door of the townhouse at street level. The fire was confined to the basement. A truck company opened the basement at grade at the rear of the structure, and a small fire was observed. A company officer at the basement door requested permission to extinguish the fire but the incident commander denied his request since he knew that crews were in the building and he did not want to have an opposing hose stream situation. The fire grew rapidly and extended up the basement stairs into the living areas of the townhouse where Firefighters Matthews, Phillips, and others were working.

With the exception of Firefighters Matthews and Phillips, all firefighters exited the building after the progress of the fire made the living area of the townhouse untenable. On the exterior of the building, firefighters realized that Firefighter Matthews was not accounted for. Firefighters re-entered the building and followed the sound of a PASS device. They removed the firefighter with the activated PASS to the exterior of the building. Once outside, firefighters realized that the firefighter who had been rescued was not Firefighter Matthews, but was, in fact, Firefighter Phillips. The search continued and Firefighter Matthews was discovered and removed approximately 4 minutes later.

Firefighter Phillips' PASS device was of the type that is automatically activated when the SCBA is activated and it worked properly. Firefighter Matthews' PASS was a manually activated type and it was found in the "off" position.

Both firefighters were burned extensively (Phillips over 60 percent of his body and Matthews over 90 percent of his body); they were transported rapidly to the hospital. Firefighter Phillips was pronounced dead upon arrival at the hospital and Firefighter Matthews died the following day. Two other firefighters were injured fighting the fire. One of the firefighters, who suffered burns over 60 percent of his body surface area, survived and was released from the hospital in August 1999. Additional information about this incident can be found in NIOSH Fire Fighter Fatality Investigation 99–F–21.

5/31/1999	**Joseph F. Tagliareni, Jr., Firefighter**	**Volunteer, Age 34**
	Secaucus Fire Department, NJ	**Heart Attack**

Firefighter Tagliareni was driving an engine in response to a report of an automobile fire. He started to feel ill, pulled the apparatus to the side of the road, exited the apparatus, walked to the rear of the engine, and collapsed from a heart attack. Other firefighters began CPR immediately and advanced life support care was provided before transporting Firefighter Tagliareni to the hospital. Firefighter Tagliareni was in a coma for 13 days before his death on 6/13/1999. The autopsy found hypertrophy and dilatation of the heart, and pneumonia.

6/2/1999	**Rudolf Cohen, Deputy Fire Chief**	**Career, Age 67**
	Gary Fire Department, IN	**Cardiac**

Deputy Chief Cohen collapsed and died of heart failure while working at his desk. No autopsy was performed.

6/3/1999	**Vincent Fowler, Captain**	**Career, Age 47**
	New York City Fire Department, NY	**Smoke Inhalation**

Captain Fowler and a probationary firefighter were searching for victims and fire in the basement of an occupied residential structure. The basement was extremely congested with furniture, newspapers, magazines, and other items. Water had been applied to the fire and it appeared to be under control. While in the basement, both firefighter low air alarms sounded. Captain Fowler delayed his own exit from the basement as he located his probationary firefighter. In the process, he ran out of air; Captain Fowler and the probationary firefighter buddy-breathed until Captain Fowler collapsed. The probationary firefighter dragged Captain Fowler until he was joined by other firefighters. Because of the congested conditions in the basement, it took an extended period of time for Captain Fowler to be removed from the building. Captain Fowler died the next day. The cause of death was listed as smoke inhalation with carbon monoxide inhalation; the level of carbon monoxide in his blood was 60 percent.

6/4/1999	**Richard Anthony Heinze, Firefighter**	**Career, Age 47**
	Newark Fire Department, NJ	**Cardiac**

Firefighter Heinze had just returned from a fire run that turned out to be a malfunctioning fire alarm system. He retired to the bunkroom shortly after returning to the station. Approximately 40 to 50 minutes after returning to the station, other firefighters heard a loud noise and found Firefighter Heinze on the floor of the bunkroom gasping for air. He soon became unconscious. Firefighters immediately began CPR and an ambulance was summoned. Despite medical care that included defibrillation, Firefighter Heinze was pronounced dead at the hospital. Firefighter Heinze had complained of pain in his jaw and a headache earlier in the shift. The cause of death was listed as hypertensive and atherosclerotic cardiovascular disease. Additional information about this incident can be found in NIOSH Fire Fighter Fatality Investigation 99–F–31.

6/8/1999	**Clyde Peterson, Assistant Fire Chief**	**Volunteer, Age 70**
	Hilltop Lakes Volunteer Fire Department, TX	**Heart Attack**

Chief Peterson had just finished commanding a fire involving a stake-bed truck filled with asphalt shingles. As he was returning to his vehicle, he suffered an apparent heart attack. He was discovered by other firefighters when they noticed that he was missing. CPR was begun immediately. No autopsy was performed. The cause of the fire was listed as suspicious.

6/9/1999   **Phillip P. Cirrito, Firefighter**   **Career, Age 47**
                Merion Fire Company of Ardmore, PA   **Pulmonary Embolism**

Firefighter Cirrito had been on duty for approximately 22 hours. During the shift, he had responded to two fire incidents and had assisted with the installation of radio equipment on a new pumper. The day was extremely hot with temperatures above 90 degrees. At approximately 5:50 a.m., an ambulance was dispatched to the fire station to a report of a firefighter gasping for air. Firefighter Cirrito died sometime later that day. The cause of death was listed as a pulmonary thromboembolism and hypertensive cardiovascular disease.

6/11/1999   **Wayne Robert Luecht, Assistant Chief**   **Career, Age 47**
               Northbrook Fire Department, IL   **Burns**

Firefighters from the Northbrook Fire Department responded to a report of an electrical problem in a department store located in a mall. Crews smelled smoke in the building and located some electrical equipment that had burned. There was no hazard of fire extension so crews were released to return to quarters, and the on-duty fire prevention staff were requested to the scene. Assistant Chief Luecht, the Fire Marshal, arrived and accompanied a private electrical contractor as he investigated the cause of the power outage and fire. As the electrician was testing an electrical panel, a white–blue flash, concussion, and fireball occurred, enveloping the electrician and Chief Luecht. Although severely burned over 90 percent of his body surface, Chief Luecht directed the response to the emergency and directed firefighters to assist other victims before he allowed them to treat him. He was alert and oriented throughout his treatment and spoke with others until he was placed on a ventilator at the hospital; however, he died on 6/21/1999. The electrician was killed and two department store employees were injured.

6/13/1999   **Arch Russell Sligar, Jr., Firefighter**   **Volunteer, Age 52**
               Bethany Volunteer Fire Department, WV   **Heart Attack**

Firefighter Sligar was riding in the officer's seat of a pumper responding to a mutual-aid structure fire. The apparatus driver noticed that Firefighter Sligar was not talking and that he had slumped over in the officer's seat. The driver stoopped the apparatus and initiated CPR. An advanced life support unit arrived within 2 minutes and defibrillated Firefighter Sligar. He regained consciousness for a short time but suffered another arrest in the ambulance en route to the hospital. He was pronounced dead about 40 minutes after arrival at the hospital. The cause of death was listed as a heart attack. No autopsy was performed. Additional information about this incident can be found in NIOSH Fire Fighter Fatality Investigation 99–F–24.

6/16/1999   **Clifford Thomas Moore, Fire Captain**   **Career, Age 38**
               Manteca Fire Department, CA   **Trauma**

Captain Moore was participating in training exercises at a regional training facility. He was proctoring the hose aloft and ladder rescue evolutions. During the second evolution, the incident commander initiated an emergency evacuation of the building as part of the training. Captain Moore attempted an emergency evacuation method from the second story window of the training tower onto a ground ladder that had been placed at the window. He fell from the window and received critical facial and head injuries. He was treated by paramedics on the scene and transported to the hospital where he was pronounced dead. The Manteca Fire Department conducted a board of inquiry into the incident. The report is available for download at the Manteca Fire Department Web site. Additional information about this incident can be found in NIOSH Fire Fighter Fatality Investigation 99–F–25.

6/17/1999   **Paul Francis McGrath, Firefighter**   **Career, Age 50**
               Pittsburgh Fire Department, PA   **Cardiac**

Firefighter McGrath was a member of a truck company that was fighting a three-alarm defensive fire in a three-story brick building that had last been used as living quarters for nursing students. Firefighter McGrath participated in numerous tasks on the fireground, including establishing a water supply to his truck company for master stream operations, ventilation, placement of ground ladders, and forcible entry. The fire had just been brought under control when the call went out that a firefighter was down. Firefighter McGrath had become dizzy at the aerial ladder turntable while operating the ladder pipe and was assisted to the ground by other firefighters. Upon reaching the ground, he collapsed of an apparent heart attack. Advanced life support treatment was provided by on-scene EMS crews, and he was transported to the hospital where he died. The cause of death was listed as arteriosclerotic cardiovascular disease. The fire was found to be arson, and suspects were arrested. Firefighter McGrath was born at the hospital located on the same grounds as the nurse's residence that burned. Additional information about this incident can be found in NIOSH Fire Fighter Fatality Investigation 99–F–22.

| 6/18/1999 | Ronald Gregory Phillips, Firefighter Paramedic | Career, Age 32 |
| | Sylvania Township Fire Department, OH | Cerebral Aneurysm |

Firefighter Phillips had just begun his daily physical fitness workout when he collapsed. He had been working out with dumbbells on a bench. His collapse was not witnessed, but others in the area heard noise and found him unconscious. ALS treatment was provided and he was transported to the hospital. Firefighter Phillips never regained consciousness and died. An autopsy was performed and the cause of death was listed as a ruptured cerebral aneurysm.

| 6/22/1999 | Wayne Rosen, Firefighter/Paramedic | Volunteer, Age 24 |
| | Akron Fire Department, NY | Trauma |

Firefighter Rosen was struck and killed by a drunk driver while riding his motorcycle to participate in training at the firehouse.

| 6/23/1999 | Matthew Eric Black, Firefighter | Volunteer, Age 20 |
| | Lakeport Fire Department, CA | Electrocution |

A large branch from a mature oak tree fell on some power lines and brought them down causing a grass fire. The Lakeport Fire Department was dispatched to the fire and warned about the downed wires. Firefighter Black's workplace was about 1-1/2 miles from the fire scene, so he responded directly to the scene in his personal vehicle and joined up with an engine company. Firefighter Black asked if he could advance a booster reel line and extinguish a pile of burning debris, and permission was granted. According to witness accounts, Firefighter Black appeared to stumble after pulling on the hose after it hung up. He fell face down on the live wire and was electrocuted. Other firefighters on the scene used a hand tool to remove the wire from under Firefighter Black and dragged him away from the wire. They initiated CPR. Advanced life support arrived within 9 minutes, and Firefighter Black was transported to the hospital where he died. Additional information about this incident can be found in NIOSH Fire Fighter Fatality Investigation 99–F–26.

| 7/4/1999 | Roger B. McEwen, Sr., Captain | Volunteer, Age 52 |
| | Hanover Volunteer Fire Department, AL | Heart Attack |

Captain McEwen had been working on the scene of a mobile home fire for about 1 hour. He had advanced and operated hose lines, helped other firefighters with their SCBAs, and was operating a fire pump when he collapsed from an apparent heart attack. Parmedics on scene began care immediately. Despite efforts on the scene and during transport, Captain McEwen was pronounced dead at the hospital. No autopsy was performed.

7/7/1999	Costello Nathaniel "Colonel" Robinson, Firefighter/	Career, Age 64
	Technician	Dog Attack
	District of Columbia Fire Department, Washington DC	

Firefighter Robinson was the most senior active firefighter on the District of Columbia Fire Department. He and his engine company were dispatched, along with other units, to the report of a fire in a densely populated area of the District. As Firefighter Robinson and other firefighters were searching for a fire, Firefighter Robinson was attacked from behind by an unrestrained pit bull terrier dog. Firefighter Robinson was injured by the attack and unable to walk. He was transported to a local hospital and scheduled for knee surgery to repair the damage caused by the dog attack. On 7/9/1999, the day that his knee surgery was scheduled, Firefighter Robinson became acutely short of breath and unresponsive. Autopsy findings included "hypertensive cardiovascular disease" and "blunt impact trauma [to the knee] with avulsion of [the] right quadriceps tendon." Additional information about this incident can be found in NIOSH Fire Fighter Fatality Investigation 99–F–40.

| 7/7/1999 | Lawrence D. Lehman, Fire Police Lieutenant
South Lebanon Township Fire Police/Friendship
  Fire Company, PA | Volunteer, Age 51
Heart Attack |

Fire Police Lieutenant Lehman was performing crowd control and traffic duties at the scene of a wildland fire. Firefighters had been on the scene for less than 1 hour when Lieutenant Lehman collapsed of an apparent heart attack. Firefighters initiated emergency medical care immediately; however, he was pronounced dead at the hospital.

| 7/12/1999 | David Vernon Parks, Firefighter/Photographer
Washington Township Volunteer Fire Department, PA | Volunteer, Age 69
Heart Attack |

Firefighter Parks was engaged in routine maintenance of an engine company apparatus when he was stricken with a heart attack and later died.

| 7/15/1999 | Bryan Christopher Pottberg, Firefighter/Paramedic
Lee's Summit Fire Department, MO | Career, Age 25
Drowning |

Firefighter/Paramedic Pottberg was participating in a scheduled rescue diver drill. Pottberg was under water performing a search evolution when he failed to surface. Other firefighters searched for him, and he was brought to the surface after approximately 11 minutes. He received immediate medical attention in the boat and while en route to the hospital, where he was pronounced dead. It is not known why Firefighter/Paramedic Pottberg had difficulty. Additional information about this incident can be found in NIOSH Fire Fighter Fatality Investigation 99–F–29.

| 7/18/1999 | Martin Michael Stiles, Inmate Firefighter
Los Angeles County Fire Department, CA | Wildland Part-Time, Age 40
Trauma |

Firefighter Stiles was a part of a Los Angeles County Fire Department Strike Team working a wildland fire incident in Ventura County. His crew was assigned to construct a handline around a slopover that extended over a dozer line on a ridge. At 1:50 a.m., Firefighter Stiles slipped over a ridge and fell to his death 150 feet below.

| 7/27/1999 | David Dwayne Hartwick, Firefighter
New Braunfels Fire & Rescue, TX | Volunteer, Age 35
Heart Attack |

Firefighter Hartwick was at home asleep when his fire department was dispatched to a structure fire. Firefighter Hartwick was found dead in the morning, the victim of a heart attack. His pager had been reset after the page for the structure fire. It is assumed that Firefighter Hartwick rose to respond to the structure fire and then died of the heart attack. This fact cannot be established with 100 percent certainty.

| 7/29/1999 | Richard F. Devine, Firefighter
Philadelphia Fire Department, PA | Career, Age 49
Heart Attack |

Firefighter Devine was assigned to an engine company working on the scene of a structure fire. He was the nozzle person and extinguished a fire on the second floor of the structure. After taking a short break once the fire was controlled, Firefighter Devine re-entered the structure to assist with overhaul. Upon his arrival at the second floor, he collapsed of an apparent heart attack. CPR was begun immediately by other firefighters and paramedics, and Firefighter Devine was transported to the hospital where he was pronounced dead. The autopsy listed the cause of death as arteriosclerotic cardiovascular disease, with heat stress cited as another significant condition. A child playing with matches caused the fire. Additional information about this incident can be found in NIOSH Fire Fighter Fatality Investigation 99–F–50.

7/30/1999 **Kenneth F. Clinch, Firefighter** Volunteer, Age 52
 Mount Marion Fire Department, NY Heart Attack

Firefighter Clinch and members of his fire department were fighting a car fire on the New York State Thruway. Firefighter Clinch had driven a water tanker to the scene and was directed to stretch a 2-1/2-inch line from the tanker to the pumper. The day was hot and humid. He stretched 100 feet of line from the tanker to the pumper, made the connection to the pumper, and was headed back to the tanker when he collapsed of an apparent heart attack. On-scene firefighters immediately went to his aid and summoned advanced life support care. The ambulance arrived within 8 minutes. Firefighter Clinch was breathing and had a weak pulse when he left the scene for the hospital; however, he died later that day. The autopsy reported the cause of death as occlusive coronary artery disease.

8/5/1999 **Richard Clarence Bacon, Assistant Chief** Volunteer, Age 46
 Dunstable Volunteer Fire Department, MA Heart Attack

Assistant Chief Bacon had driven a fire department pumper to the scene of a reported structure fire. The fire was minor, and all units were ordered to return to the station. Upon returning to the station, Chief Bacon was completing some paperwork when he was observed to be in medical distress. A firefighter in the station began CPR, and he was transported to the hospital where he died.

8/5/1999 **James Everett Clark, III, Senior Firefighter** Career, Age 42
 Midwest City Fire Department, OK Struck by Vehicle

Senior Firefighter Clark was a member of a squad company that had been dispatched to the report of a motor vehicle collision on I–40 in Midwest City. The roads were wet from rain and rain had begun to fall again. A ladder company also was dispatched on the call. The squad arrived on the scene and discovered that the collision was minor. The ladder arrived and positioned itself behind the squad to divert traffic away from the incident scene; all of the unit's emergency lights were operating. Approximately 2 minutes after arriving on the scene, the ladder was hit from behind by a passenger vehicle. Firefighters dismounted the ladder apparatus to check on the condition of the driver. Senior Firefighter Clark, who had heard the collision, joined the firefighters. After the patient from the second collision was moved to an area that was thought to be safe (between the ladder truck and the wall), the company officer of the ladder company walked further upstream in traffic in an attempt to wave traffic away from the scene of both collisions. At this point, another passenger vehicle lost control and spun into the space between the ladder apparatus and the retaining wall. Senior Firefighter Clark placed himself between the oncoming car and the driver of the car that had collided with the ladder apparatus. Two firefighters and the driver of the car that hit the ladder apparatus were injured. Senior Firefighter Clark died as the result of head injuries on 8/8/1999. Additional information about this incident can be found in NIOSH Fire Fighter Fatality Investigation 99–F–27.

8/5/1999 **Michael Eugene "Cuppie" Cupp, Sr., Fire Chief** Volunteer, Age 48
 Cygnet Volunteer Fire Department, OH Heart Attack

Chief Cupp and members of his fire department had just extinguished a brush fire on a vacant lot. As Chief Cupp returned to his truck, he collapsed of an apparent heart attack. He was pronounced dead later that night at the hospital. The cause of death was a massive myocardial infarction. No autopsy was performed.

8/5/1999 **Cilton Jay Dauzat, Firefighter** Volunteer, Age 63
 White Tail Ridge Volunteer Fire Department, TX Struck by Apparatus

Firefighter Dauzat had responded with members of his fire department to a wildland fire that resulted from the failure of a local homeowner to contain an intentional fire in a pile of logs. The driver of the responding pumper did not realize that Firefighter Dauzat had mounted the back step. The pumper was attempting to ascend a hill. Despite two tries, the pumper was unable to climb the hill and when the apparatus backed down off of the hill, the driver discovered the he had run over Firefighter Dauzat. It is believed that Firefighter Dauzat lost his footing. The cause of death was listed as severe chest and head injuries.

8/8/1999   **Ronald Wade Meshell, Firefighter**   **Volunteer, Age 30**
                        Huttig Fire Department, AR   **Burns**

Firefighter Meshell responded, along with others from his department, to a mutual-aid call for a fire in a motor home. The owner of the motor home had just filled the fuel tank, driven the motor home to his residence, and parked the vehicle at the top of his driveway. Upon arriving home, the owner noticed smoke coming from under the hood; he could not extinguish the fire and called the fire department. Firefighter Meshell and a deputy fire chief arrived in a pumper and parked about 40 feet downhill from the motor home. Firefighter Meshell was ordered to pull a 1-1/2-inch line from the rear of the Huttig pumper to assist a firefighter from another department that had a booster line from the other department's pumper on the fire. Before the line could be charged, the fuel tank on the motor home ruptured and sent a flood of burning fuel down hill toward Firefighter Meshell, another firefighter, and both pumpers. Firefighter Meshell was not wearing any protective clothing, although he had loaded his protective clothing on the pumper prior to response. He was surrounded by flames for an estimated 15 seconds and was burned over 96 percent of his body surface and his airway. He died on 8/16/1999. Additional information about this incident can be found in NIOSH Fire Fighter Fatality Investigation 99–F–34.

8/9/99   **Arthur J. Heckman, Firefighter**   **Volunteer, Age 66**
                        Macks Creek Fire Department, MO   **Heart Attack**

Firefighter Heckman was the first to arrive at a wildland fire that was threatening structures. Firefighter Heckman had pulled a pumper behind a house to fill it with water. He was found slumped over the wheel, having suffered an apparent heart attack.

8/14/1999   **Frank William Wood, Firefighter**   **Volunteer, Age 54**
                        Flourtown Fire Company, PA   **Heart Attack**

Firefighter Wood and members of his department were fighting a room fire in a nursing home. Firefighter Wood assisted with the rescue of an unconscious staff member and then collapsed of an apparent heart attack. He was rushed to the hospital but subsequently died.

8/26/1999   **David Thomas Nall, Assistant Chief**   **Volunteer, Age 40**
                        Town of Jay Volunteer Fire Department, FL   **Heart Attack**

Assistant Chief Nall responded to an emergency medical incident on a high school football field. Chief Nall assisted with the treatment of the patient and was helping to load the child in an ambulance when he complained of chest pains. He collapsed and CPR was initiated. Chief Nall was transported to a local hospital where he was pronounced dead about 1 hour later. No autopsy was performed.

8/27/1999   **Stephen Joseph Masto, Firefighter**   **Career, Age 28**
                        Santa Barbara Fire Department, CA   **Heat Stroke**

Firefighter Masto was working as an EMT at a wildland fire, roaming among other firefighters and providing first aid to anyone who became injured. He worked a 6 a.m. to 6 p.m. shift, was equipped with a portable radio, and carried a canteen. He did not return to camp at the end of his shift, and a search was initiated. Firefighter Masto was discovered dead, wearing brush gear, about 12 hours later (on 8/28/1999) in steep terrain. The cause of death was found to be heat stroke; there was no evidence of trauma or other medical conditions that contributed to his death.

8/31/1999   **Timmy Roger Dawson, Firefighter**   **Volunteer, Age 34**
                        Center Rock Volunteer Fire Department, SC   **Apparatus Rollover**

Firefighter Dawson was the driver of a 1994 pumper responding to a motor vehicle collision. Two other firefighters were passengers in the pumper. The right wheels of the pumper left the road, and Firefighter Dawson attempted unsuccessfully to bring the truck back under control. He overcompensated and the pumper went off the left side of the road, through a yard, and rolled several times. The pumper's speed was estimated at 60 miles per hour in a 35-mile-per-hour zone. Firefighter Dawson was not wearing a seat belt. The other firefighters riding on the pumper received minor injuries. The cause of death was listed as blunt trauma.

9/13/1999	**Kenneth C. Cashman, Firefighter**	**Volunteer, Age 29**
	Auglaize Township Volunteer Fire Department, OH	**MVA**

Firefighter Cashman was responding to a stove fire in a residence. A dump truck loaded with stone pulled out in front of Firefighter Cashman's car, colliding with it and killing him instantly. Firefighter Cashman was wearing a seat belt at the time of the collision and was operating a red dash light and a siren. The dump truck driver was charged with aggravated vehicular homicide.

9/15/1999	**Terri LeAnn Hood, Firefighter**	**Volunteer, Age 31**
	McColloch Volunteer Fire Department, IN	**Struck by Apparatus**

Firefighter Hood was helping firefighters from a five-county area battle a 450-acre wildland fire. She was protecting a tobacco barn with other firefighters when conditions worsened and firefighters decided to withdraw. In heavy smoke and confusing conditions, Firefighter Hood was run over by a pumper backing away from the barn. She was killed instantly. Additional information about this incident can be found in NIOSH Fire Fighter Fatality Investigation 99–F–35.

9/27/1999	**Lewis Edward "Rawhide" Anderson, Firefighter**	**Volunteer, Age 68**
	River Falls Volunteer Fire Department, SC	**Struck by Vehicle**

Firefighter Anderson was directing traffic around an earlier traffic collision. He was positioned about a 1/2 mile north of the collision scene. The weather was rainy and Firefighter Anderson was wearing bright yellow rain gear and using a stop/slow sign. He was struck by an 18-wheel truck, which also struck a fire department vehicle on the scene. The truck then left the scene. Firefighters were notified by another truck driver that Firefighter Anderson was down. They went to his aid and administered CPR and other medical care. He died on 9/30/1999. According to the police report, the truck driver was operating his vehicle at a reckless speed. According to the certificate of death, Firefighter Anderson was killed by a cerebral contusion and edema due to blunt force trauma to the head.

10/3/1999	**Gregory Edwin Pacheco, Firefighter**	**Wildland Part-Time, Age 20**
	United States Forest Service, Carson National Forest, NM	**Trauma**

Firefighter Pacheco was a member of a forest firefighting crew constructing a fire line on the La Jolla fire near San Diego, California. He was ascending steep terrain when a large rock fell and hit Firefighter Pacheco on the head, injuring him severely. One other firefighter received moderate injuries and was released back to his crew. Firefighter Pacheco died on 10/5/1999. The cause of death was listed as a closed head injury.

10/4/1999	**Jeffrey Scott Thompson, Firefighter**	**Volunteer, Age 20**
	Howell County Rural Fire District #1, MO	**Electrocution**

Firefighter Thompson and other members of his fire department responded to a grass fire that was caused when some power lines fell. The line had fallen on a fence, energizing it. The fire chief warned all firefighters that fences in the area were energized and to avoid them. Firefighter Thompson and two other firefighters were advancing a booster line to control the fire when they came in contact with a fence. All three were electrocuted and injured seriously. Firefighters on the scene provided medical care, and the injured were evacuated by air. Firefighter Thompson died that day. The two other firefighters survived their injuries.

10/5/1999	**William Malcolm Bethune, Captain**	**Career, Age 58**
	Texas City Fire Department, TX	**MVA**

Captain Bethune was riding in the officer's seat of an engine company responding with lights and siren to a medical emergency. As the engine company entered an intersection against the red light, it struck a passenger car, veered off the roadway, and struck a cement freeway support column. Captain Bethune, who was not wearing a seat belt, was ejected through the front windshield of the pumper. He struck the pavement and received severe injuries. Captain Bethune was provided with medical care on the scene and flown to a trauma center. He was pronounced dead upon arrival at the trauma center. The cause of death was listed as blunt trauma to the head. The driver of the pumper also was severely injured; the firefighter riding in the back of the cab sustained only minor injuries. A police investigation of the incident attributed the cause of the accident to the failure of the passenger car to yield to a responding emergency vehicle. The driver of the passenger car acknowledged that he had seen the responding engine approaching but thought that he could get through the intersection before the engine got there. The report also concluded that Captain Bethune's failure to wear a seat belt was a major factor in his death. Captain Bethune was the first firefighter fatality for the Texas City Fire Department since most members of the Texas City Fire Department were killed in an explosion 52 years before. Additional information about this incident can be found in NIOSH Fire Fighter Fatality Investigation 99–F–36.

10/7/1999	**Marvin Huisman, First Assistant Chief**	**Volunteer, Age 56**
	Wilmont Fire Department, MN	**Heart Attack**

Chief Huisman and other firefighters were battling a brush fire. Chief Huisman had assisted with the extension of hose lines and was operating a fire pump. He suffered a heart attack and died.

10/7/1999	**Elvis Benson Maxwell, Firefighter/Operator**	**Volunteer, Age 49**
	Grant Parish Fire District #5, LA	**Apparatus Rollover**

Firefighter/Operator Maxwell was responding as the driver of a fire department tanker tender to a structure fire on a rainy night. He lost control of the vehicle that left the roadway and overturned. An EMS unit en route to the fire came upon the scene and discovered Firefighter Maxwell still inside the vehicle with no vital signs. The paramedics initiated care, but he was pronounced dead in the emergency room. He was not wearing a seat belt and was partially ejected in the collision. The cause of death was listed as blunt trauma.

10/16/1999	**Karen Jane Savage, Firefighter/EMT**	**Volunteer, Age 44**
	Hawkins Bar Volunteer Fire Department, CA	**Struck by Apparatus**

Firefighter/EMT Savage and other members of her department responded to a wildland fire that had developed into firestorm conditions. The fire eventually consumed 26,000 acres and destroyed 75 homes. Firefighter Savage and other firefighters stopped at a support vehicle to get supplies. As she handed supplies to other firefighters on a pumper, the vehicle began to move. She fell or jumped from the pumper and was crushed by the pumper's rear wheels. Additional information about this incident can be found in NIOSH Fire Fighter Fatality Investigation 99–F–42.

10/18/1999	**Charles C. Young, Firefighter/EMT**	**Volunteer, Age 77**
	Ross Township Fire Department, OH	**Heart Attack**

Firefighter/EMT Young responded to a very upsetting incident involving the suicide of a teen. He returned from the incident and was speaking with his wife on the phone about the call when he was stricken with a heart attack. His wife was unsure of where he was so it took firefighters 10 to 20 minutes to find and treat him. Despite efforts by members of his department and others, Firefighter/EMT Young died.

10/28/1999	**Brian K. Burnett, Firefighter**	**Volunteer, Age 23**
	Robert Charles Ulrich, Captain	**Volunteer, Age 57**
	Scipio Township Volunteer Fire Department, IN	**Apparatus Rollover**

Firefighter Burnett and Captain Ulrich were responding in a tanker (tender) to the report of a brush fire. Firefighter Burnett was driving but failed to negotiate a curve in the road; the apparatus left the road and crossed into a cornfield, where it rolled several times. Firefighter Burnett was ejected from the vehicle and the vehicle rolled on top of him. Captain Ulrich was trapped in the tanker, which was on its roof, until he was extricated by other firefighters. Both firefighters were transported to the hospital.

Captain Ulrich died on 11/4/1999. He had been released from the intensive care unit to a regular hospital floor. Captain Ulrich was seemingly well and recovering from his injuries. He was discovered pulseless and nonresponsive; emergency care was provided but was not successful. The autopsy concluded that Captain Ulrich died of a cardiac arrhythmia. It is not known if the cardiac problems were related to the collision. Captain Ulrich was wearing his seat belt at the time of the collision.

Firefighter Burnett died on 1/22/2000. He was making a slow recovery. The family had been told that he might be home in a week or so but that he would need further therapy. No cause of death for Firefighter Burnett is available. He was not wearing his seat belt at the time of the collision.

| 10/29/1999 | **David Merle Pack, Forestry Aide I**
 Tennessee Department of Agriculture/Forestry Division, TN | **Wildland Full-Time, Age 63**
 Drowning |

Forestry Aide Pack responded to what was described as a "routine" woodland fire. Other firefighters spoke with him as they responded to the incident. At the conclusion of the incident, Forestry Aide Pack's pickup truck was found at the edge of a pond near the fire area with the headlights on and the engine running. After a foot search failed to locate him, a search dog was called in. The dog led searchers to Forestry Aide Pack's body in the pond. His body was recovered. The reason for his presence in the pond is unknown. His cause of death was drowning.

| 10/29/1999 | **Walter F. Vaughan, Fire Police Officer**
 Warminster Fire Department, PA | **Volunteer, Age 80**
 Struck by Vehicle |

Fire Police Officer Vaughan was directing traffic around the scene of a reported structure fire. He was struck by a passenger car and sustained multiple injuries. He was transported to the hospital where he was placed on a ventilator. Officer Vaughan died on 11/13/1999. At the incident, he was wearing a reflective safety vest and helmet and using a wand-type flashlight to direct traffic. The driver of the passenger vehicle was cited for careless driving and failure to obey an authorized person directing traffic.

| 11/2/1999 | **Michael J. Sims, Sr., Firefighter**
 Highland Hose Company, Tarentum, PA | **Volunteer, Age 38**
 Trauma |

Firefighter Sims was responding to an automatic fire alarm activation as a passenger in a 1965 open cab aerial ladder apparatus. As the truck made a turn, Firefighter Sims fell from the vehicle and sustained severe injuries. His fall was not witnessed, but another firefighter heard the impact of Firefighter Sims striking the pavement. Emergency medical care was provided, and he was airlifted to a hospital. Firefighter Sims died the following day. The apparatus was equipped with seat belts and a safety gate; it is not known if Firefighter Sims was wearing his seat belt at the time of his fall.

| 11/3/1999 | **Jerry Wayne Ramey, Firefighter Trainee**
 West Fork Fire Department, AR | **Volunteer, Age 18**
 Electrocution |

Firefighter Trainee Ramey responded with members of his department to a fire in the utility easement behind a home. A very small fire was found, which was out except for a few burning embers. Firefighter Trainee Ramey attempted to stomp out the embers when he came into contact with a 7,200-volt electrical line that had been hidden from view in tall grass. He fell on top of the line and was removed by other firefighters. Emergency medical aid was provided, and he was transported to the hospital where he was pronounced dead.

| 11/4/1999 | **William Walter Korte, Firefighter**
 Southampton Fire Department, NY | **Volunteer, Age 59**
 Heart Attack |

Firefighter Korte was performing scene safety duties as other members of his department performed a vehicle extrication. He had just finished closing a road for traffic control. He was standing by the fire police truck talking to another firefighter when he dropped to the ground. Firefighter Korte died of an apparent heart attack. No autopsy was performed.

| 11/7/1999 | David Zan Lancaster, Firefighter | Volunteer, Age 24 |
| | Elliott Volunteer Fire Department, MS | MVA |

Firefighter Lancaster was killed as the result of a motor vehicle collision in his personal vehicle while responding to a car fire.

| 11/14/1999 | Bert Andrew Bruecher, Firefighter | Volunteer, Age 46 |
| | Village of Pleak Volunteer Fire Department, TX | Apparatus Rollover |

Firefighter Bruecher was the driver and lone occupant of a tanker that was responding to a fire involving 200 round bales of hay that were near a home and a propane tank. The tanker entered a curve at high speed, left the road, and rolled over. A shift in the water load may have contributed to the collision. Firefighter Bruecher was partially ejected and was pinned under the truck. Two 14-year-old boys were arrested and one was charged with second-degree arson for setting the fire. As a part of a plea bargain, the boys were placed on probation until they are 18.

| 11/16/1999 | Brian Andrew Lee, Firefighter | Career, Age 38 |
| | Fire Department Jersey City, NJ | Cardiac |

Firefighter Lee and his fire company had just returned from an emergency response and were resupplying their engine company. Firefighter Lee began to experience severe stomach pains. An ambulance was called and he was loaded for transport to the hospital. While en route to the hospital, he collapsed into cardiac arrest and was not revived. The cause of death was listed as natural inflammation of the heart caused by sarcoidosis.

| 11/18/1999 | Henri Fred Broussard, Fire Chief | Volunteer, Age 69 |
| | Maurice Volunteer Fire Department, LA | Heart Attack |

Maurice firefighters responded to a fire that involved the cab of an 8,600-gallon gasoline tanker next to several above ground fuel tanks and a large liquid petroleum gas tank at a local gas station. Chief Broussard drove the first engine to the scene and was met there by other firefighters. As the firefighters dressed in their protective clothing, Chief Broussard stretched the initial attack line and then returned to the pumper to operate the pump from the top-mounted pump panel. Shortly after arriving back at the pumper, Chief Broussard suffered a heart attack. Other firefighters and the crew of an on-scene ambulance provided immediate medical aid. Chief Broussard was transported to a local hospital where he was pronounced dead upon arrival. No autopsy was performed. Three other firefighters were injured at the incident.

| 11/18/1999 | James Melvin Dunham, Safety Officer/Firefighter | Volunteer, Age 36 |
| | Saint Jo Fire Department, TX | Cardiac |

Firefighter Dunham drove a fire department rescue truck to the scene of a mutual-aid vehicle collision that required extrication. As he was setting up the power unit for a hydraulic rescue tool, Firefighter Dunham stumbled and hit his head on the ground. Other firefighters rendered immediate aid and defibrillated him. Advanced life support aid was provided on the scene, and Firefighter Dunham was flown to a hospital by helicopter. Despite all efforts, he did not survive. The cause of death was listed as occlusive coronary artery atherosclerosis.

| 11/20/1999 | Jackie Mac Garnett, Firefighter | Volunteer, Age 54 |
| | Quapaw Volunteer Fire Department, OK | Heart Attack |

Firefighter Garnett and his department were dispatched to a wildland fire. Firefighter Garnett rose to respond and suffered a heart attack. He was transported to a local hospital, stabilized, and then flown to a regional hospital. He was taken to surgery and died a short time later. No autopsy was performed.

| 11/20/1999 | Alton L. "Al" Lewis, First Assistant Fire Chief | Volunteer, Age 55 |
| | Montour Falls Fire Department, NY | Heart Attack |

First Assistant Fire Chief Lewis had just returned home after responding to a motor vehicle accident. He suffered a heart attack in his driveway and died.

11/20/1999   **Wayne C. Yost, Assistant Fire Chief**   **Volunteer, Age 48**
                 Cochranville Fire Company, PA   **Heart Attack**

Assistant Chief Yost had responded to a shed and wildland fire. He complained of not feeling well at the scene and went home. Shortly after his arrival at home, an ambulance was called. Assistant Chief Yost suffered a heart attack. He was revived at his home by members of a local response team but died on 11/27/1999.

12/2/1999   **Brad A. Michener, Firefighter**   **Volunteer, Age 24**
                 Scipio–Republic Fire Department, OH   **Heart Attack**

Firefighter Michener had driven a heavy rescue truck to the scene of a brush fire and then back to the station at the conclusion of the incident. Firefighter Michener left the station in his personal vehicle and returned home. His home was about a block from the fire station. As he exited his vehicle at home, he collapsed in his backyard of an apparent heart attack and died the next day. Firefighter Michener had a pre-existing heart condition but had been released to full duty by his personal physician.

12/3/1999   **Paul Arthur Brotherton, Firefighter**   **Career, Age 41**
                 Timothy Paul Jackson, Firefighter   **Career, Age 51**
                 Joseph T. McGuirk, Firefighter   **Career, Age 38**
                 Jeremiah Michael Lucey, Firefighter   **Career, Age 38**
                 James Francis Lyons, Firefighter   **Career, Age 34**
                 Thomas Edward Spencer, Lieutenant   **Career, Age 42**
                 Worcester Fire Department, MA   **Trapped**

Members of the Worcester Fire Department responded to a fire in the Worcester Cold Storage Warehouse. The building was a windowless six-story structure. Upon arrival, firefighters found a large warehouse with light smoke conditions and a fire on the second floor. They initiated search-and-rescue and fire-attack operations. Within seconds, conditions in the fire building changed and thick black smoke reduced visibility to zero. All fire department personnel were ordered down from upper floors and a head count was taken. It revealed that two firefighters were not accounted for. A "mayday" radio transmission was received from Firefighter Brotherton indicating that he and Firefighter Lucey were lost and running out of air. A search for the trapped firefighters was initiated with 18 firefighters searching for the two that were lost. Lieutenant Spencer, Firefighter Jackson, Firefighter McGuirk, and Firefighter Lyons entered the fifth floor to conduct a search. Contact with the teams was lost and all six firefighters died. NIOSH conducted a review of this incident. The cause of the fire was accidental, the result of a candle knocked over during a domestic dispute by transients living in the building.

12/7/1999   **Roy Kenneth Crago, Firefighter**   **Volunteer, Age 65**
                 Fallston Volunteer Fire and Ambulance   **Brain Aneurysm**
                   Company, Inc., MD

Firefighter Crago and members of his fire department were dispatched to a report of a transformer explosion. Firefighter Crago arrived at the fire station in his personal vehicle and responded as the driver of an engine company apparatus. Two other firefighters were on board. At some point during the response, Firefighter Crago suffered a stroke. The engine left the road, skidded down an embankment, and crashed into a concrete culvert. Other firefighters removed Firefighter Crago from the apparatus and began CPR and ALS care. He was transported to the hospital. Despite all efforts, he died on 12/10/1999. The cause of death was listed as a subarachnoid hemorrhage caused by a brain aneurysm. Firefighter Crago was not wearing his seat belt, but the injuries he received as a result of the crash were not life threatening.

12/10/1999   **Richard L. Van Wert, Fire Chief**   **Volunteer, Age 58**
                 Schaghticoke Fire Department, NY   **Explosion**

Chief Van Wert was supervising the disposal of fireworks residue in a controlled burn at a local fairground. He noticed a spark heading toward a van containing at least 100 pounds of additional residue that was to be discarded. He yelled for the fireworks company employee to run but was unable to escape the explosion himself. The van exploded and burned. Chief Van Wert was killed instantly. His actions were credited with saving the employee.

12/13/1999	**Gregory Eugene Rodgers, Firefighter/EMT**	**Volunteer, Age 50**
	Dresden Volunteer Fire Department, OH	**Heart Attack**

Firefighter Rodgers responded as the passenger of a water tanker that responded to a mutual-aid barn fire. The driver of the tanker was his son, a firefighter/EMT. Firefighter Rodgers assisted with the setup of a portable tank and suction equipment and helped the tanker dump its load of water at the fire scene. Once empty, the tanker responded to a hydrant about 2 miles from the fire scene and connected to it. The driver of the tanker found Firefighter Rodgers on the ground and unresponsive. He summoned paramedics and began CPR. Firefighter Rodgers was transported to the hospital where he died.

12/15/1999	**Paul Franklin Ezernack, Jr., Firefighter**	**Volunteer, Age 28**
	North Sabine Fire Protection District, LA	**Apparatus Rollover**

Firefighter Ezernack was responding in a 1,500-gallon tanker to a report of a brush fire. An embankment gave way under the right wheels of the tanker. He attempted to regain control, but the tanker left the roadway and rolled over. He was ejected from the vehicle and thrown 170 yards. He was pronounced dead at the scene.

12/18/1999	**Bradley Curtis McNeer, Firefighter**	**Volunteer, Age 22**
	Chesterfield County Fire Department, VA	**MVA**

Firefighter McNeer was riding in the officer's seat of a heavy rescue responding to a gas leak in a residence. He and the driver were the only occupants of the vehicle. Neither firefighter was familiar with the route to the incident, and Firefighter McNeer was having difficulty finding the address in the apparatus map book. The driver decided to reduce his response mode to non-emergency and pull over to look at the map book himself. As he was preparing to pull over or stop, the right rear wheels of the apparatus left the road and went into a ditch. The driver was able to steer the unit out of the ditch, but oversteered to the left and struck a car. He then saw another oncoming vehicle and overcorrected to the right ending up back in the ditch. The rescue truck struck a large tree. Firefighter McNeer was wearing a seat belt but received a fatal head injury. A fire department investigation concluded, "The accident was caused by the driver taking his eyes off the road as he reached for the light switch. Attempting to drive the vehicle out of the ditch, and the speed of the vehicle, contributed to the severity of this accident."

12/20/1999	**John H. Tvedten, Battalion Chief**	**Career, Age 47**
	Kansas City Fire Department, MO	**Trapped**

Chief Tvedten was a sector officer working inside a warehouse that was involved in fire. Visibility in the warehouse was good, and firefighters were putting water on the fire. About 45 minutes into the incident, Interior conditions changed rapidly as thick black smoke enveloped the building. The incident commander ordered the building to be evacuated, and Chief Tvedten ordered firefighters to leave. The emergency evacuation signal was given over radios and by fire apparatus air horns at the scene. During the evacuation, Chief Tvedten became disoriented and lost. He was in radio communication with Command. Six search teams swept the building but were not able to locate the chief.

12/20/1999	**Theodore A. Ferrante Jr., Firefighter**	**Career, Age 43**
	Revere Fire Department, MA	**Heart Attack**

Firefighter Ferrante and members of his ladder crew responded to an activated fire alarm in a highrise building. The cause of the alarm was found to be a prank pull station activation. At the scene of the incident, Firefighter Ferrante complained of chest pains but told his company officer that he just needed to lie down and he would feel better. Approximately 2 hours later, Firefighter Ferrante began to experience severe pain. An ambulance was called as other firefighters rendered aid. Firefighter Ferrante was transported to a local hospital where he died.

12/22/1999	Jason L. Bitting, Firefighter	Career, Age 29
	David M. McNally, Assistant Chief	Career, Age 48
	Nathan R. Tuck, Firefighter	Career, Age 39
	Keokuk Fire Department, IA	Trapped

The Keokuk Fire Department was dispatched to a fire in a residential structure. The structure was a house built in 1910 that was divided into three apartments. The department responded with an engine, a quint, and a chief's vehicle with a total of three firefighters, a lieutenant, an assistant chief, and the fire chief. The response of the chief and one firefighter was delayed because they were returning from a previous incident. Upon arrival, Assistant Chief McNally, Firefighter Bitting, and Firefighter Tuck entered the building in full turnouts and SCBA for search and rescue. A mother and child were trapped on the roof above the porch and three other children were trapped inside. Firefighters rescued one infant child who was transported to the hospital by a police officer. Firefighters rescued a second infant child who was transported to the hospital by a police captain and the fire chief. The fire chief was away from the scene for approximately 3 minutes. Firefighters were searching for the third child when a flashover occurred and trapped all three. An aggressive fire attack was mounted by firefighters who were arriving as part of a callback of off-duty members but the effort was not able to save the lives of the three firefighters. Assistant Chief McNally was found on the second floor at the top of the stairs with the third child. Firefighter Bitting was found in the front bedroom on the second floor of the apartment. Firefighter Tuck was found on the first floor of the apartment in the living room area. All three firefighters were wearing PASS devices that were integrated with their SCBAs. The PASS devices failed to sound an alarm when the firefighters became incapacitated. The SCBAs and PASS devices are undergoing testing to determine why they did not operate.

In addition to the three firefighters killed in this incident, the two infant children and a 7-year-old child perished. A child playing with the stove caused the fire; two high chair trays that were stored on top of the stove caused the fire. Smoke alarms in the home did not operate.

| 12/29/1999 | Ronald Eugene Kaltreider, Safety Officer | Volunteer, Age 39 |
| | Pleasant Hill Volunteer Fire Company, PA | Heart Attack |

Safety Officer Kaltreider had been at the fire station for most of the day performing year-end computer work and assuring that his department was prepared for Y2K. As he discussed the upcoming purchase of some communications equipment with another firefighter, he suffered a heart attack. Firefighters immediately began attempts to revive him, but they were unsuccessful. Officer Kaltreider had a history of heart disease dating back to 1990.

| 12/31/1999 | Robert Dale Pollard, Firefighter | Volunteer, Age 64 |
| | Southern Stone County Fire Protection District, MO | Stroke/CVA |

Firefighter Pollard was driving a rescue vehicle to a wildland fire. While en route, he collapsed and was treated by first responders and then airlifted to a hospital. Firefighter Pollard died the next day of a cerebral bleed (stroke/CVA).

| 1/8/2000 | Lee A. Purdy, Pump Operator/Inspector | Volunteer, Age 57 |
| | Spencerville Invincible Fire Company, OH | Heart Attack |

Pump Operator/Inspector Purdy was operating a top-mounted pump panel at the scene of a residential structure fire. Inspector Purdy asked his wife, a volunteer paramedic, for a drink. When she returned to the truck with the drink, she saw him fall from the truck, the victim of a massive heart attack. Firefighters provided medical aid immediately. Inspector Purdy was transported to a local hospital where he was pronounced dead 20 minutes after arrival. No autopsy was performed.

| 1/11/2000 | Ronald J. Osadacz, First Assistant Chief | Volunteer, Age 36 |
| | Morganville Volunteer Fire Company Number One, NJ | Heart Attack |

First Assistant Chief Osadacz was on the scene of a vehicle fire that resulted from the collision of a pickup truck with a tree. While working on the scene, he was struck in the groin area by a water stream from a one and 1/2-inch hoseline. He was agitated by this occurrence, left the scene, and returned to his home. Upon his arrival at home, Chief Osadacz complained of indigestion, took some over-the-counter medicine, and laid down to rest. Within a few moments, he suffered a fatal heart attack. An autopsy revealed that he died of severe occlusive coronary arteriosclerosis. A physician who examined him less than a week prior to his death stated that the chief's agitated state would have contributed to the heart attack.

1/11/2000	**Allen L. Streeter, Firefighter**	**Volunteer, Age 58**
	Ranch Drive Fire District, OK	**Heart Attack**

Firefighter Streeter responded to a trash and grass fire in the department's brush truck. Shortly after exiting the vehicle, he collapsed of an apparent heart attack. Firefighters immediately initiated CPR, and an ambulance was summoned.

1/12/2000	**Robert M. Jones, Firefighter**	**Volunteer, Age 48**
	Unity Volunteer Fire Department, ME	**Heart Attack**

Firefighter Jones was preparing to use a dry hydrant to supply water for a fire in a residential structure. He attached a large diameter hoseline to the pumper, removed the cap from the dry hydrant, and was preparing to attach a suction hose to the hydrant. Another firefighter, who was assisting Firefighter Jones, came around the truck and found him on the ground. After calling for help, the assisting firefighter began CPR. Despite treatment on the scene and in the ambulance, Firefighter Jones died. The cause of death was listed as a heart attack. The fire was caused when a boy and his two brothers inadvertently set a sofa on fire; all three boys, 6-year-old triplets, died in the fire.

1/15/2000	**Gary Lynn Bankert, Sr., Firefighter**	**Volunteer, Age 37**
	Roanoke–Wildwood Volunteer Fire Department, NC	**Drowning**

Firefighter Bankert was participating in fire department sponsored dive training in a rock quarry that contained a private lake used exclusively for recreational diving. Firefighter Bankert was a member of his department's search and recovery dive team. As the class ascended from the third of three dives, the class stopped for a safety and accountability check at a depth of 15 feet. At the time of the check, Firefighter Bankert was present; however, when the class proceeded to the surface, he did not surface. Other divers went immediately to the bottom of the lake and found Firefighter Bankert at a depth of approximately 22 feet. He was brought to the surface and transported by paramedic ambulance to a local hospital. He was pronounced dead later that evening. The cause of death was listed as severe metabolic acidosis as the result of a near drowning. Additional information about this incident can be found in NIOSH Fire Fighter Fatality Investigation F2000–11.

1/16/2000	**Ernest John Young, Firefighter/Trustee**	**Volunteer, Age 52**
	Big Knob Volunteer Fire Department, Station 26, PA	**Trauma**

Firefighter/Trustee Young was assisting with the replacement of electric garage door openers on apparatus bay doors at his fire station. He and another firefighter had climbed to the top of a fire rescue truck using a 14-foot extension ladder. As Firefighter Young began his descent, the ladder slipped out from under him and he fell approximately 10 feet and struck his head on the concrete floor. The ladder was not being footed at the time it fell. Despite immediate medical aid and transport by helicopter to a regional hospital, Firefighter Young died on 1/17/2000. The cause of death was listed as blunt force trauma to the head. Additional information about this incident can be found in NIOSH Fire Fighter Fatality Investigation F2000–07.

1/17/2000	**James William Altic, Fire Chief**	**Volunteer, Age 47**
	Halfway Fire & Rescue, MO	**Apparatus Rollover**

Chief Altic was the lone occupant and driver of a tanker apparatus responding with lights and siren to a mutual-aid structure fire. Road conditions were slippery because a light misty rain was falling after prolonged dry spell. Chief Altic failed to negotiate a curve in the road, and the apparatus left the roadway and rolled over. He sustained fatal neck and chest injuries and was pronounced dead at the scene. The driver's seat belt had been removed from the apparatus at some point prior to the collision, so Chief Altic was not restrained. The police report cited the speed of the fire apparatus as a factor in the collision, as well as the wet roadway. Additional information about this incident can be found in NIOSH Fire Fighter Fatality Investigation F2000–18.

1/17/2000	**Juan Gilberto De Leon, Captain**	**Career, Age 53**
	McAllen Fire Department, TX	**Heart Attack**

Captain De Leon was in his assigned sector driving a command vehicle. He stopped at a business in his sector to help a civilian move some boxes. During this task, Captain De Leon suffered a heart attack. The civilian called 9–1–1 and provided CPR until the arrival of fire department and EMS responders. Additional information about this incident can be found in NIOSH Fire Fighter Fatality Investigation F2000–12.

1/27/2000	**Walter Harvey Gass, Captain**	**Volunteer, Age 74**
	Sealy Volunteer Fire Department, TX	**Smoke Inhalation/Burns**

Captain Gass and other members of his department were dispatched to a residential structure fire that was caused when lightning struck a house. The first two firefighters on the scene, the assistant chief and the fire chief, confirmed a working fire with dark smoke and fire visible from the attic and dormers. Captain Gass and his crew were the first fire company to arrive at the scene. He and two other firefighters entered the structure through the front door to perform an attack on the fire. Shortly afterwards, as the two firefighters who were with Captain Gass fed more hose into the structure, there was a rapid buildup of heat and the hoseline seemed to drop. The firefighters exited the building and reported the situation to the chief. Two rapid intervention teams were formed and, after four attempts, the second team was successful in recovering Captain Gass. He was equipped with full structural protective clothing and a manually activated PASS device. The PASS was found in the "off" position. Captain Gass was located about 18 feet inside the front door of the structure. Firefighters removed him from the structure approximately 20 minutes after his arrival on the scene. The cause of death was listed as smoke and soot inhalation with greater than 80 percent total thermal injury. Additional information about this incident can be found in NIOSH Fire Fighter Fatality Investigation F2000–09.

1/27/2000	**Robert Boy Ketelsen, Firefighter**	**Volunteer, Age 59**
	Westbrook Fire Department, CT	**Heart Attack**

Firefighter Ketelsen and members of his fire department responded to an automatic fire alarm. The alarm turned out to be unfounded. Fire department members returned to the fire station and placed the fire apparatus in service. Less than 2 minutes after departing the fire station for home in his personal vehicle, Firefighter Ketelsen suffered a heart attack. He managed to pull off the road into a parking lot before he became unconscious. Firefighter Ketelsen was found in full cardiac arrest when members of his fire department arrived. He was transported to a local hospital where he was pronounced dead. Firefighter Ketelsen had a history of heart problems.

2/6/2000	**Douglas George Stevens, Training Officer**	**Volunteer, Age 42**
	Story City LaFayette Township Volunteer Fire Department, IA	**Heart Attack**

Training Officer Stevens was working on the scene of a residential structure fire. He climbed a ground ladder, used a halligan tool to remove a section of siding, and then continued on to the roof to prepare to perform additional ventilation. Training Officer Stevens descended the ladder and walked toward a backup hose team that was standing by outside of the residence. As he neared the other firefighters, Training Officer Stevens collapsed due to an apparent heart attack. Paramedics standing by on the scene initiated care immediately. ALS medical care was provided during a 17-minute transport to the hospital to no avail. The cause of death was listed as occlusive coronary artery disease. The cause of the fire was an overheated wall next to a chimney. Additional information about this incident can be found in NIOSH Fire Fighter Fatality Investigation F2000–14.

| 2/11/2000 | Paul Eugene Cooper, Firefighter | Volunteer, Age 26 |
| | Hoopa Volunteer Fire Department, CA | MVA |

Firefighter Cooper was responding as the driver of an engine apparatus en route to a motor vehicle accident on a narrow, two-lane road. While the engine was about to negotiate a slight left curve, a car approached from the other direction straddling the line between the two lanes. Firefighter Cooper moved the apparatus to the right side of the road to avoid a collision, and the engine's right tires left the pavement and drove onto a soft grassy shoulder. The truck continued on the shoulder, began to fishtail, glanced off of a power pole on the right side of the road, veered to the left out of control, and struck a large oak tree. Neither Firefighter Cooper nor the passenger was wearing a seat belt. Firefighter Cooper was trapped behind the steering wheel, it took firefighters nearly 1 hour to free him. The other firefighter, a passenger, was ejected. An ambulance transported both firefighters to the hospital. Although he was alert and conscious during the extrication, Firefighter Cooper entered a coma in the hospital. He never regained consciousness and died on 2/14/2000, his 27th birthday.

| 2/13/2000 | Richard Owen Spink, Lieutenant | Career, Age 48 |
| | Fort Campbell Fire Department, KY | Heart Attack |

Lieutenant Spink had just completed participating in a live burn structural training session. Lieutenant Spink was participating in a critique of the training when he was struck with a massive heart attack. Lieutenant Spink had a number of cardiac risk factors including prior heart attacks or chest pain, high blood pressure, overweight, diabetes, and smoking.

2/14/2000	Lewis Evans Mayo, III, Firefighter	Career, Age 44
	Kimberly Ann Smith, Firefighter	Career, Age 30
	Houston Fire Department, TX	Trapped

Firefighters Mayo and Smith responded with Engine Company 76, three other engines, two ladder companies, two chief officers, an ambulance, and support staff to a report of a fire in a McDonald's restaurant. The fire was reported at 4:30 a.m. Engine Company 76 was staffed with a captain, a fire apparatus operator, and two firefighters. Engine 76 was the first firefighting unit, arriving at 4:38 a.m., and reported 6-foot flames visible from the roof; the flames appeared as if they might be venting from an exhaust fan, possibly indicating a grease fire. Firefighters from Engine 76 and other units were ordered to advance an attack line into the interior of the structure where no fire was visible. At 4:52 a.m., the incident commander ordered all firefighters out of the building to transition to a defensive fire attack. The flames visible from the roof had grown to 30 feet in height, and fire had become visible in the kitchen area of the restaurant.

After the evacuation order was given, Firefighters Mayo and Smith were declared missing. The incident commander requested a second alarm at 5:02 a.m. and numerous rescue attempts were made. At 5:27 a.m., the third alarm was requested. Shortly afterwards, a ladder company opened the rear door of the restaurant and made entry to the back of the kitchen area. A PASS device had been heard alarming in the kitchen area, and a firefighter was able to see a downed firefighter through the door. Firefighter Mayo was discovered with his facepiece in place, his regulator not connected to the facepiece, and with his SCBA partially removed and entangled in wires. He was removed and transported to the hospital, where he was pronounced dead.

Given the amount of time that had passed and the likelihood that Firefighter Smith was buried in debris, the search effort transitioned into a recovery mode. Firefighter Smith was found at approximately 7:13 a.m. within 6 feet of the rear door of the restaurant. She was entangled in wires and a pair of wire cutters was found near her body. She was wearing an SCBA, but the status of her facepiece and regulator could not be determined. Both firefighters died of asphyxia due to smoke inhalation. Firefighter Mayo's carboxyhemoglobin level was found to be 26 percent and the level for Firefighter Smith was found to be 52 percent.

The fire was intentionally set by a group of juveniles attempting to conceal a burglary attempt. Four individuals were convicted of crimes with sentences ranging from 2 to 35 years. Additional information about this incident can be found in NIOSH Fire Fighter Fatality Investigation F2000–13.

2/19/2000　　**James D. Geiger, Firefighter/EMT**　　　　**Volunteer, Part Paid, Age 55**
　　　　　　　　City of Defiance Fire & Rescue, OH　　　　**Heart Attack**

Firefighter/EMT Geiger responded with other members of his fire department to a sledding accident. Firefighter/EMT Geiger assisted with patient packaging and helped carry the patient through deep snow to an ambulance. Firefighter/EMT Geiger left the scene at the conclusion of the incident in his private vehicle. Firefighter/EMT Geiger suffered a heart attack as he arrived at his home; his car struck a propane tank. Firefighters were called to the scene and transported Firefighter/EMT Geiger to the hospital where he died shortly after arrival. The autopsy cited severe occlusive coronary artery disease as the cause of death.

2/21/2000　　**Evangelino Soto Rodriguez, Sergeant**　　　**Career, Age 53**
　　　　　　　　Puerto Rico Fire Department, PR　　　　　**Struck by Vehicle**

Sergeant Rodriguez was called to the scene of an arson-caused lumberyard fire. Sergeant Rodriguez had just finished attaching a hoseline to a hydrant and began to cross the road back to his engine company. As he crossed the road, a car operated by a drunk driver struck him. The cause of death was listed as severe multiple trauma.

2/29/2000　　**Robert Jeffery Jackson, Firefighter**　　　**Volunteer, Age 35**
　　　　　　　　Harmony Volunteer Fire Department, OK　　　**MVA**

Firefighter Jackson was responding to a mutual-aid grass fire in a neighboring community. Firefighter Jackson was the sole occupant and driver of a 3/4-ton, 4-by-4 brush truck. As Firefighter Jackson responded, he encountered a sedan traveling in the opposite direction. As the sedan crested a hill, the driver lost control, skidded approximately 258 feet, crossed the center of the road, and struck the brush truck head on. Firefighter Jackson attempted to avoid the collision by pulling to the side of the road. The emergency lights on the brush truck were activated. Firefighter Jackson was wearing a seat belt. Firefighter Jackson's speed was estimated at 30–35 miles per hour and the speed of the sedan was estimated at 70 miles per hour.

After the collision, both vehicles caught fire. The fire was reported to the Harmony Volunteer Fire Department, and they responded to the incident. Upon arrival, both vehicles were found to be fully involved in fire. The cause of death was listed as massive blunt chest trauma with burns noted as another significant medical condition.

3/4/2000　　**David Paul Sutton, Firefighter**　　　　**Volunteer, Age 27**
　　　　　　　Fraser Department of Public Safety, MI　　　**Asphyxiation**

Firefighter Sutton responded, along with other members of his department, to a working apartment fire. While they were engaged in suppression of the first fire, another apartment fire was reported in a building across the street from the original fire. Since no fire apparatus was available to respond, Firefighter Sutton and other firefighters responded in a van to the scene. Police officers were in the process of evacuating the building. A resident in need of rescue had been spotted at a second-story window. Mutual-aid fire companies were responding but not yet on the scene. The smoke condition at the entrance to the apartment building was light, with heavier smoke and heat on the second floor. One firefighter observed fire at the top of the stairs. Firefighter Sutton and another firefighter, equipped with full protective clothing and SCBA, entered the building to effect the rescue. Witnesses outside the building reported that the resident disappeared from the window as if she had been reached by firefighters. Within seconds, a flashover occurred, trapping the resident and the two firefighters. Both firefighters managed to reach a bathroom at the rear of the apartment, but they were unable to get through the window with their SCBAs in place. Firefighter Sutton was observed by other firefighters at the window, and a rescue effort was mounted.

Two firefighters shed their SCBA and entered the bathroom from ground ladders. Firefighter Sutton was removed after his SCBA was cut from him. The low-pressure hose on his SCBA had burned through. The other firefighter was located in the bathtub and removed. Both were transported to the hospital where Firefighter Sutton was pronounced dead. The other firefighter sustained major burns and was hospitalized for 6 months; the apartment resident died.

The fire was caused when an arsonist(s) ignited combustibles on the first and second floors of the apartment building. This fire was one of six arson fires that occurred in the same general area over 2 days.

3/6/2000	**Robert W. Buhler, Firefighter**	**Volunteer, Age 62**
	Delmont Volunteer Fire Department, SD	**Overrun by Wildfire**

Firefighter Buhler and members of his fire department were fighting a wildland fire. The fire was the result of a controlled field burn that was being conducted by some local citizens that got out of control. The conditions were dry with winds of 40 miles per hour. The fire was in a very deep winding ravine. Hose was being added to an attack line when a wind gust blew up an area that had been thought to be previously extinguished. The fire spread rapidly up a hill and engulfed Firefighter Buhler. Firefighter Buhler had responded directly to the scene from a nearby town and was not wearing protective clothing. Firefighter Buhler was severely burned over 60 to 80 percent of his body and died on 3/16/2000. Another firefighter, who was near Firefighter Buhler at the time of the blowup and who was equipped with protective clothing, received minor injuries. Additional information about this incident can be found in NIOSH Fire Fighter Fatality Investigation F2000-22.

3/6/2000	**Donald R. Wilson, Assistant Chief**	**Volunteer, Age 50**
	Herrick Fire Protection District, IL	**Heart Attack**

Assistant Chief Wilson was on the roof of a residence that was involved in fire. The fire started when a garbage fire extended through brush to a house. He had just chopped a hole in the roof to determine if the fire had spread to the attic. He was seen lying on the roof. Firefighters found that Assistant Chief Wilson had suffered a heart attack. Despite their efforts, Assistant Chief Wilson was pronounced dead at the scene.

3/7/2000	**Jerry Wayne Coppin, Training Officer**	**Volunteer, Age 56**
	Okay Volunteer Fire Department, OK	**Cerebral Hemorrhage**

Training Officer Coppin and members of his department responded to assist with storm watch duties in their community. Toward the end of the storm, Training Officer Coppin suffered a cerebral hemorrhage as he sat in his pickup truck and operated a radio. First responders and ambulance personnel provided medical aid immediately. Training Officer Coppin was transported to a hospital but died on 3/11/2000.

3/8/2000	**William M. Blakemore, Private**	**Career, Age 48**
	Javier Lerma, Lieutenant	**Career, Age 41**
	Memphis Fire Department, TN	**Shot**

Along with other units, Engine 55, a four-person engine company including Lieutenant Lerma and Private Blakemore, responded to the report of a residential structure fire. Engine 55 was the first unit on the scene and reported a working house fire. Lieutenant Lerma stepped from the apparatus to perform a size up of the fire and was immediately shot by a gunman who had been hiding in the garage of the house. The gunman continued to fire, striking Private Blakemore as he sat in the back of the pumper preparing his protective equipment. The driver of Engine 55 moved the apparatus forward out of the danger zone. The fourth member of the crew had been on the opposite side of the apparatus preparing his protective clothing and jumped onto the running board as the apparatus was moved to safety. As the gunman engaged and killed a deputy sheriff, firefighters moved Lieutenant Lerma to a safe area and began treatment.

Lieutenant Lerma was pronounced dead at the scene and Private Blakemore died in an ambulance en route to the hospital. Both were killed by shotgun blasts to the head. A deputy sheriff was also killed. After the scene was secured, firefighters extinguished the fire in the house and discovered the body of the gunman's wife. The gunman was an off-duty Memphis firefighter. He was responsible for the fire in the house, and he was likely the person who reported the fire to the Memphis Fire Department through a 9–1–1 call.

Lieutenant Lerma's father was killed in the line of duty in 1977. Lieutenant Lerma was carried to his rest in a fire truck named after his father.

3/13/2000	Jessie Lamar Y'Barbo, Forestry Technician III	Wildland Career, Age 54
	Texas Forest Service, TX	Burns

Forestry Technician Y'Barbo was participating in the controlled (prescribed) burn of a 35-acre block of mature pine forest, with low understory vegetation. Forestry Technician Y'Barbo was wearing brush gear and operating a 1985 Honda 250 all-terrain vehicle (ATV), owned by the Forestry Service. The ATV had a shop-made holder at the rear that accommodated a drip torch. As Forestry Technician Y'Barbo ascended an incline, the vehicle overturned backward. As he struggled to free himself, he accidentally kicked off the cap on the ATV's fuel tank, splashing him with fuel, which was ignited by the drip torch. Other firefighters quickly came to Forestry Technician Y'Barbo's aid, and he was transported to a hospital where he died on 4/7/2000. The cause of death was listed as complications of thermal injuries.

An Accident Review and Mitigation Report prepared by the Forest Service recommended that all Forest Service ATVs be equipped with threaded caps (the cap on Forestry Technician Y'Barbo's ATV was only equipped with a 1/4-turn cap); fuel tank venting and overflow tubes; operator training, equipping supervisors with first aid kits that include fire blankets; the consideration of replacement of older, narrow track ATVs; and the installation of rollover protection on ATVs.

3/15/2000	Mike Shortt, Fire Chief	Career, Age 44
	Weaverville Volunteer Fire Department, CA	Heart Attack

Chief Shortt was acting as an instructor for an evening drill on ground ladder placement and raises. Chief Shortt acted mainly as an instructor/coach and did not personally participate in much actual ladder handling. During the class, Chief Shortt collapsed, the victim of a heart attack. Medical aid was provided immediately by firefighters attending the class. Chief Shortt was transported to the hospital where he died 3/31/2000.

3/17/2000	David Clements Sharp, II, Firefighter/Engineer	Career, Age 31
	Fayetteville Fire/Emergency Management Department, NC	Train Collision

Firefighter/Engineer Sharp responded to an automatic fire alarm as the driver and lone occupant of a 1993 Pierce Arrow 100-foot ladder tower truck. The first unit on the scene found a system malfunction and canceled all other responding fire apparatus, including the ladder truck operated by Firefighter/Engineer Sharp.

As he returned to the fire station, Firefighter Sharp came upon a railroad crossing that was blocked by traffic control devices as a slow-moving freight train passed. As the last car of the train passed, it stopped just past the intersection. Firefighter Sharp drove the ladder truck around the traffic control arm and attempted to cross the tracks. As he passed the freight train, a passenger train, traveling approximately 30 mph, headed in the opposite direction of the freight train struck the left front of the fire truck. Witness statements indicated that Firefighter/Engineer Sharp' view of the oncoming passenger train was likely blocked by the freight train. The collision spun the truck around and he was ejected; he landed under the truck's rear tires and was pronounced dead at the scene. The cause of death was listed as multiple blunt trauma. He was not wearing a seat belt. Additional information about this incident can be found in NIOSH Fire Fighter Fatality Investigation F2000–19.

3/18/2000	Frederick L. Brain, Fire Police Officer	Volunteer, Age 76
	Miller Place Fire Department, NY	Heart Attack

Fire Police Officer Brain was directing traffic at the scene of a motor vehicle collision. After approximately 1 hour on the scene, Fire Police Officer Brain collapsed. The fire chief reached

Fire Police Officer Brain and found him gasping for breath and having a weak pulse. Firefighters on the scene applied an automatic external defibrillator and the unit delivered a shock. A firefighter performed CPR on Fire Police Officer Brain as an ambulance was summoned. Fire Police Officer Brain was transported to the hospital with a pulse. He lapsed into a coma and died on 4/24/2000.

3/27/2000	**Kevin Francis Sterenchuk, Administrative District Chief**	**Career, Age 48**
	Cedar Rapids Fire Department, IA	**Heart Attack**

District Chief Sterenchuk was in his office performing administrative duties. He had not been feeling well an hour prior and had asked the department's EMS coordinator to take his blood pressure. His blood pressure was high, and he was also complaining of an ache in his right elbow similar to carpal tunnel pain. He agreed to have his blood pressure checked again. Later that day, the EMS coordinator walked past District Chief Sterenchuk's office and noticed that he was having difficulty breathing. He directed a staff member to call 9–1–1 and retrieved medical equipment from a reserve fire apparatus stored in the building. Revival efforts were unsuccessful.

3/28/2000	**Michael "Mike" Russell Queen, Fire Chief**	**Volunteer, Age 30**
	Clayton Fire Department, GA	**Trauma**

Fire Chief Queen was assisting with hose testing at his fire station. During the testing process, a 2-1/2-inch hose separated from its coupling. High-pressure water struck Chief Queen and propelled him into a fire truck that was parked nearby. Chief Queen suffered a fatal blow to the head as he hit the apparatus. A firefighter/EMT began treatment immediately, and Chief Queen was transported to the county hospital. He was pronounced dead approximately 45 minutes after the accident.

3/31/2000	**Kendall O. Bryant, Firefighter/EMT**	**Paid-on-Call, Age 36**
	Layton City Fire Department, UT	**Smoke Inhalation**

Firefighter/EMT Bryant and members of his department were dispatched to the report of a residential fire. Upon arrival, Firefighter/EMT Bryant's captain reported a working fire with flames and smoke visible from the garage. The captain ordered his firefighters to extinguish the fire in the garage, and the fire was knocked down within 5 minutes of their arrival on scene. The captain then instructed Firefighter/EMT Bryant and another firefighter to enter the structure with a hoseline to search for victims, conduct fire extension, and begin ventilating the structure. Upon entry, the firefighters encountered dark smoke but no visible flame. They began a left-hand search and proceeded to the second floor of the structure. The second floor contained bedrooms and was directly above the garage. A lieutenant joined the firefighters on the second floor by following the hoseline. As the firefighters searched the bedrooms, there was a rapid buildup of heat. A red glow was visible at the bottom of the stairs, cutting off the team's escape route. The decision was made to follow the hoseline back out of the structure since the firefighters were unsure about the presence of windows in the bedrooms and the stairway was small. Firefighter/EMT Bryant was the last in line as the firefighters made their way to safety.

As the firefighters emerged from the house, the lieutenant removed his facepiece and told other firefighters that Firefighter/EMT Bryant was supposed to be right behind him but had not exited the structure with him. The incident commander ordered an accountability check and Firefighter/EMT Bryant was confirmed to be missing. A second crew of firefighters entered the residence through the front door but could not climb the stairs because they appeared to be collapsed and were heavily involved in fire. The incident commander ordered a ladder raised to provide firefighters with access to a roof area, which led to the bedroom windows. Two firefighters entered the second floor of the structure and searched two bedrooms. A sound believed to be Firefighter/EMT Bryant's PASS device was located but turned out to be a smoke alarm.

The firefighters saw a light in the bedroom across the hall and found that it was a flashlight that was carried by Firefighter/EMT Bryant. Firefighter/EMT Bryant was found on his knees on the floor with his facepiece removed. His SCBA cylinder was found to be empty and his protective hood was found over his mouth and nose, most likely in an attempt to filter air to breathe. His PASS device was found in the "off" position. Firefighter/EMT Bryant was removed by firefighters through a window and lowered to the ground into the care of waiting paramedics.

Firefighter/EMT Bryant was pronounced dead in the emergency room. The cause of death was later listed as smoke and soot inhalation and acute carbon monoxide poisoning. Firefighter/EMT Bryant's blood carboxyhemoglobin level was found to be 25 percent at the time of his death. Firefighter/EMT Bryant was a career firefighter in Ogden, Utah. Two other firefighters were injured. The fire was caused by a droplight that had been hung near a cardboard box that was being used as part of a dog's bed. Additional information about this incident can be found in NIOSH Fire Fighter Fatality Investigation F2000–23.

| 4/4/2000 | **David Anthony Maisano, Captain** | **Volunteer, Age 38** |
| | **Tritown Fire Department, MI** | **Pneumonia** |

Captain Maisano and members of his fire department responded to a report of smoke in a residential structure. After removing the contents of the fireplace and checking the attic and roof of the structure, it was determined that smoke was escaping the chimney in the attic and causing the smoke buildup. Firefighters were stowing equipment in preparation for their return to the station. Captain Maisano was attaching an elastic cord to secure a ground ladder to the truck when the cord snapped and struck him in the face. Captain Maisano fell to the ground from a height of approximately 9 feet, fracturing a wrist and causing severe back pain. He was transported to the hospital by ambulance; he was in the hospital for 3 days prior to being discharged. Due to his pain he was unable to sleep in bed, so he slept in an easy chair on the night that he was released from the hospital. He was last seen alive at approximately 2:30 a.m. on April 8th, but was discovered dead by his wife in the morning. The autopsy attributed his death to pneumonia. The autopsy also noted that the use of multiple prescribed pain relief medications might have resulted in some degree of respiratory depression.

| 4/7/2000 | **James Ted Griffith, Firefighter/Training Officer** | **Volunteer, Age 25** |
| | **Winterset Volunteer Fire Department, IA** | **Explosion** |

Firefighter Griffith and members of his department were called to the scene of a grass fire. The fire started when salvage workers ignited grass and nearby wood as they worked to dismantle two old, rusting 12,000-gallon elevated fuel storage tanks. The grass fire was extinguished, and the salvage workers decided to use a blowtorch to cut a small hole near the drain of the tank that had already been pulled to the ground. The hole was intended to allow the attachment of a tow chain, which would be used to pull the tank to a salvage yard. As the hole was being made with a blowtorch, the tank emitted a hissing sound and suddenly exploded. Firefighter Griffith was killed instantly when he was struck by flying debris. The top of the tank, which was torn away in the explosion, weighed over 900 pounds and flew over 114 feet before coming to rest. A salvage worker was also killed, and eight firefighters and a civilian received injuries. Analysis of the tank contents revealed that the tank contained residual gasoline and other petroleum products.

| 4/11/2000 | **Michael R. Baughn, Firefighter** | **Paid on Call, Age 46** |
| | **Washington Fire/Rescue, OH** | **Cardiac** |

Firefighter Baughn and 30 other members of his department were participating in search-and-rescue training in the basement of an office building. Teams of two firefighters, equipped with structural protective clothing and SCBA, were doing crawl-through searches in areas that were obscured with nontoxic smoke. After completing an exercise that lasted approximately 20 minutes, Firefighter Baughn sat in a basement hallway, removed his SCBA facepiece, rested, and waited for other firefighters to complete the exercise. When the other firefighters emerged from the exercise, Firefighter Baughn began to climb the stairs from the basement to the ground level of the building. As he reached the first landing in the stairwell, Firefighter Baughn suddenly collapsed. Firefighters found Firefighter Baughn and immediately initiated CPR.

Arriving ambulance personnel were unable to deliver a shock to Firefighter Baughn through their defibrillator. Firefighter Baughn was transported immediately to the closest hospital, 2 minutes away. Firefighter Baughn was pronounced dead 30 minutes later.

The case of death was listed as cardiomegaly (enlarged heart)—acute cardiac arrhythmia. Firefighter Baughn had a history of high blood pressure. He took two medications for this disease, but he did not take them regularly. The energy selector for the defibrillator was found to be in the "0" mode. The defibrillator will not deliver a shock unless an energy level greater than zero for the shock is selected. Additional defibrillator training was recommended for squad members. Additional information about this incident can be found in NIOSH Fire Fighter Fatality Investigation F2000–24.

4/20/2000	**Rickey Levi Davis, Firefighter/Paramedic**	**Career, Age 33**
	Center Point Fire/Rescue, AL	**Hyperthermia**

Firefighter/Paramedic Davis and members of his department were dispatched to a report of a fire in a single-family residential structure that included a full basement. Upon arrival, firefighters found heavy smoke showing from the structure and found that the fire was in the basement. Firefighters attempted to reach the fire through the garage door (which opened into the basement) but were unsuccessful in locating the seat of the fire. A positive-pressure fan was placed at the garage door. Another team of three firefighters, including Firefighter/Paramedic Davis, advanced an attack line through the front door of the residence. On their initial entry into the residence, they were unable to locate any fire. The crew withdrew, found that a positive-pressure fan had been placed at the front door, and returned to explore another area of the house.

Firefighter/Paramedic Davis was at the nozzle as the hoseline was advanced into the second entry on the main floor of the residence. As the line was advanced, Firefighter/Paramedic Davis fell through the floor into the area of the basement that was involved in fire. Other firefighters helped Firefighter/Paramedic Davis as he attempted to jump back to the first floor from the basement but these efforts were unsuccessful. Firefighters attempted to lower a scuttle hole ladder into the hole but the location of the hole and the sagging of the first floor into the basement prevented its use. Firefighters instructed Firefighter/Paramedic Davis to use the hoseline to protect himself as they attempted to rescue him through the basement.

An attack team entered the basement and fought its way to the room that contained Firefighter/Paramedic Davis. It is estimated that 12 to 15 minutes passed from the time Davis fell into the basement until he was located and removed from the structure. Upon removal from the structure, he was flown to the hospital where he was pronounced dead. The cause of death was listed as hyperthermia (thermal injuries). The carboxyhemoglobin level in Firefighter/Paramedic Davis' blood was less then 5 percent; he was burned over roughly one-third of his body.

4/26/2000	**Robert Cowey Brannon, Jr., Lieutenant**	**Career, Age 43**
	Bluefield Fire Department, WV	**Cardiac**

Lieutenant Brannon responded with his three person engine company to a report of a residential structure fire. Arriving companies found a working fire on the first and second floors of a 1-1/2-story wood-frame house. Lieutenant Brannon took command of the incident and ordered an attack line through the front door. After completing a 360-degree walk around the involved structure, Lieutenant Brannon assisted with the deployment and advancement of a second attack line. The line was stretched to the first floor of the house and was used to control hot spots. Lieutenant Brannon noticed an air leak on his SCBA. The leak was controlled by another firefighter inside the house, but Lieutenant Brannon found that he was out of air and needed a new cylinder. Lieutenant Brannon exited the house, spoke momentarily with the fire chief, and proceeded to his apparatus to get his cylinder changed. He kneeled at the truck to allow the driver/operator to replace his cylinder. The driver/operator asked Lieutenant Brannon if he was okay, Lieutenant Brannon responded that he needed a new cylinder, turned his head, and collapsed.

Other firefighters and on-scene paramedics immediately came to Lieutenant Brannon's aid. He was defibrillated several times on the scene and transported to a local hospital. He was revived in the emergency room but died on 5/4/2000. The cause of death was listed as atherosclerotic coronary artery disease. Additional information about this incident can be found in NIOSH Fire Fighter Fatality Investigation F2000–34.

4/29/2000	David John Liston, Smokejumper	Wildland Part-Time, Age 28
	Bureau of Land Management/Alaska Fire Service, AK	Parachute Failure

Smokejumper Liston was participating in mandatory annual recertification practice parachute jumps in preparation for the upcoming wildland season. He had completed three jumps. During the fourth jump of the day, Smokejumper Liston's parachute failed, and he plunged 3,000 feet to his death. Emergency medical care was provided immediately by other smokejumpers trained as emergency medical technicians to no avail. His cause of death was listed as multiple impact (deceleration) injuries. Smokejumper Liston was the first parachute-related fatality for the Bureau of Land Management in 40 years; all smokejumping activity in Alaska and Idaho was halted for over 2 months as this incident was investigated.

The investigation revealed that the parachute malfunction was characterized as a "drogue in tow" meaning that the drogue chute deployed but did not release on demand to deploy Smoke-jumper Liston's main parachute. Smokejumper Liston then followed emergency procedures and manually deployed his reserve parachute. During this action, the reserve pilot chute became entangled with the drogue bridle (the line which attaches the drogue to the main parachute) thereby preventing both the main and reserve canopies from deploying.

4/29/2000	L.C. Merrell, Lieutenant	Career, Age 43
	Chicago Fire Department, IL	MVA

Lieutenant Merrill was in command of a truck company responding with lights and siren to a still alarm. Lieutenant Merrill was riding in the front passenger seat of the apparatus and was not wearing a seat belt. The truck company slowed prior to entering an intersection and was broad-sided by a pickup truck that ran a stop sign. Lieutenant Merrill was ejected. Despite immediate medical care, he was pronounced dead at the scene. The cause of death was listed as blunt head trauma. Four other firefighters and nine civilians in two vehicles were injured. The still alarm turned out to be false. The Chicago Fire Department Commissioner was quoted as saying that Lieutenant Merrell could have survived the accident if he had been wearing a seat belt. The driver of the pickup was ticketed for speeding and failure to stop.

4/30/2000	Arnold Blankenship, III, Second Assistant Chief	Volunteer, Age 27
	Greenwood Volunteer Fire Company #1, Inc., DE	Trapped

Assistant Chief Blankenship and other members of his department were participating in a train-ing/demolition burn of a 2-1/2-story, wood-frame dwelling. According to the fire chief, the plan for the day was to do small, one-room burns to evaluate a saw, and then to completely burn the house. After a series of small fires were extinguished on the first floor of the house, preparations were made for the demolition burn.

The plan for the final fire was to ignite the attic, then ignite the first floor, evacuate the house, and allow it to burn completely. Water curtain nozzles were set up on the exterior of the house to protect trees that were in the proximity of the house. Assistant Chief Blankenship went into the attic of the house and used a small garden-type sprayer to distribute diesel fuel in the attic. As fires were ignited inside an attic room, Assistant Chief Blankenship used the sprayer to "accelerate" the fire. With the exception of Assistant Chief Blankenship, all firefighters had left the attic space and were proceeding to the first floor of the structure. A firefighter waiting at the base of the attic stairs for Assistant Chief Blankenship noted fire and smoke coming from the attic. When firefighters reached the exterior of the structure, they notified the fire chief that Assistant Chief Blankenship was missing and possibly trapped.

As some firefighters attempted to suppress the fire, other firefighters used a ground ladder to access the second floor of the house in an attempt to rescue Assistant Chief Blankenship. After several attempts, firefighters followed the sound of an activated PASS device and were able to reach Assistant Chief Blankenship. They were unable to remove him as portions of the collapsed roof covered him. mutual-aid firefighters arrived and were able to locate and remove Assistant Chief Blankenship's body about an hour after the time he was reported missing. The cause of death was later listed as asphyxiation and burns.

5/7/2000	**Carl Ray Payne, Pilot**	**Wildland Part-Time, Age 66**
	Payne Flying Service, United States Department of	**Aircraft Crash**
	the Interior for the Texas Forest Service, TX	

Pilot Payne had just dropped a load of fire retardant on a fire outside of Fort Stockton, Texas. The aircraft was an Air Tractor AT–802A. After the drop, Pilot Payne was circling two radio antennas (280 feet and 310 feet) when the outboard 5 feet of his right wing struck the antenna guide cables and support structure. Pilot Payne was able to level the aircraft and continued to fly; however, the aircraft was compromised and struck trees and terrain before coming to rest. Cable strands from the antenna towers were found among the aircraft wreckage. The cause of death for Pilot Payne was listed as multiple severe trauma. Additional information related to this incident may be found in National Transportation Safety Board Accident Investigation FTW00TA142.

5/8/2000	**Kenneth Jesse, Fire Police Officer**	**Volunteer, Age 80**
	Harford Volunteer Fire Company, PA	**Heart Attack**

Fire Police Officer Jesse had responded with members of his department to a vehicle fire on the interstate. On the scene of the vehicle fire, he blocked traffic to protect firefighters engaged in control of the fire. When the fire was extinguished, Fire Police Officer Jesse returned home. Upon his arrival at home, Fire Police Officer Jesse told his wife that he felt dizzy and had a lump in his stomach. His wife went inside their house for a moment to prepare to take him to the doctor and found that he had collapsed in their carport when she returned outside. His wife called 9–1–1, and firefighters from Fire Police Officer Jesse's department responded and began CPR. Despite their efforts, Fire Police Officer Jesse died of a heart attack.

5/15/2000	**Leo Koponen, Air Attack Pilot**	**Wildland Contract Part-Time,**
	Courtney Aviation, United States Forest Service, CA	**Age 49**
	Samuel James Tobias, Air Tactical Group Supervisor	**Wildland Career, Age 47**
	United States Forest Service, Lincoln National	**Aircraft Crash**
	Forest, NM	

Air Attack Pilot Koponen and Air Tactical Group Supervisor Tobias were beginning a reconnaissance flight to look for any fires that may have spread from the ongoing Scott–Able fire. The aircraft was a Cessna T337C. Approximately 6 minutes after takeoff, black smoke was noted in the vicinity and calls were received from local residents reporting smoke in Alamo Canyon. A helicopter flying in the area confirmed that there was a downed aircraft and that there appeared to be no survivors. Investigators found that the aircraft had crashed nose first and that the death of both firefighters was immediate. The weather at the time of the flight was clear with a light wind. Additional information related to this incident may be found in National Transportation Safety Board Accident Investigation DEN00GA089.

5/27/2000	**Evan N. Shirk, Firefighter**	**Volunteer, Age 27**
	Moreau Fire Protection District, MO	**Apparatus Rollover**

Firefighter Shirk was the sole occupant and driver of an engine apparatus returning from a motor vehicle accident. The pumper was equipped with a 1,000-gallon water tank that was filled to 900 gallons. The accident turned out to be an unfounded report. As Firefighter Shirk returned to the station, the right front wheel drifted off of the road onto soft ground. According to the police accident report, Firefighter Shirk overcorrected and the right wheel struck a drainage culvert, causing the pumper to veer across the road, roll over several times, and catch fire. Firefighter Shirk was not wearing a seat belt and was ejected. Firefighter Shirk was pronounced dead at the scene. The cause of death was listed as massive head trauma. The police accident report listed driver inattention as a factor in the crash. Additional information about this incident can be found in NIOSH Fire Fighter Fatality Investigation F2000–33.

5/31/2000	**Lyndell J. Smith, Firefighter**	**Volunteer, Age 46**
	Caldwell Fire Department, KS	**Apparatus Rollover**

Firefighter Smith was a passenger in a 1984 Jeep CJ7 command vehicle. The truck was equipped with extrication equipment and was responding with lights and siren in operation to a vehicle rollover with reports of serious injuries. Firefighter Smith was riding the hump between the driver and the front seat passenger. None of the truck's occupants were wearing seat belts. As the command vehicle overtook and passed a passenger car on the left side, the car turned into the command vehicle, striking it at the right rear tire. The command vehicle skidded across traffic, entered a ditch, overturned in a wheat field, and caught fire. All three occupants of the command vehicle were ejected. Firefighter Smith received fatal injuries; the other occupants were seriously injured. No autopsy was performed but the cause of death for Firefighter Smith was listed as closed head and chest trauma with exsanguination from mortal axillary wounds (blood loss). No extrication was required at the scene of the original rollover call.

6/4//2000	**George A. "Bo" Burton, Firefighter/Rotor Craft Pilot**	**Wildland Career, Age 48**
	Florida Division of Forestry, FL	**Helicopter Crash**

Firefighter Burton was fighting a fire near Fort Myers, Florida. He had been on the scene for approximately 1-1/2 hours, performing reconnaissance, making water drops, and filling his external bucket from a local lake. Witnesses observed the helicopter in level flight headed back to the lake after a water drop. The helicopter was reported to suddenly bank deeply with its nose down. After a few seconds, the helicopter crashed in a cow pasture. The cause of the crash has not been determined. The aircraft was a 1966 Bell UH1 (205). Additional information related to this incident may be found in National Transportation Safety Board Accident Investigation MIA00GA184.

6/25/2000	**Whitney C. Teehan, Jr., Captain**	**Volunteer, Age 66**
	Eastern Point Volunteer Fire Company #2, CT	**Heart Attack**

Captain Teehan and members of his department responded to a manufacturing facility for a report of fire. Captain Teehan was acting as the department's Accountability Officer on the scene of the incident. It was determined that a large dust cloud caused by a high-pressure air leak had been mistaken for smoke. As companies prepared to return to quarters, Captain Teehan suffered a massive heart attack while seated in a pumper. Other firefighters began CPR and an AED was used. The AED was unable to restore a rhythm, and Captain Teehan was transported to the hospital, where he died. Captain Teehan had suffered a previous heart attack on the scene of a car fire in 1997. An AED operated by a private fire department saved him that day.

7/2/2000	**Nathan Andrew Pescatore, Firefighter**	**Volunteer, Age 17**
	Lloydsville Volunteer Fire Department and Relief	**MVA**
	Association, PA	

Firefighter Pescatore was responding as the sole occupant and driver of his personal vehicle to a report of a structure fire. He crossed the centerline of the road as he entered a curve in the road. As he rounded the curve, he came upon a farm tractor approaching from the opposite direction. Firefighter Pescatore's view of the tractor as he drove into the curve was blocked by vegetation.

Firefighter Pescatore was unable to get back into his lane and struck the farm tractor head on. The loader bucket on the front of the tractor was driven through both driver's side roof posts and severely injured Firefighter Pescatore. Firefighters responding on mutual aid to the structure fire were diverted to the collision and were joined by Lloydsville firefighters at the scene. After Firefighter Pescatore was extricated, he was flown by helicopter to the hospital, where he was pronounced dead.

7/15/2000   **Phillip Ridings, Firefighter**   **Volunteer, Age 52**
                      Hornersville Volunteer Fire Department, MO   **Heart Attack**

Firefighter Ridings was advancing a 1-1/2-inch hose line at the scene of an electrical fire in a two-story residence. The home was originally built in 1917 and had been renovated several times. As he worked, Firefighter Ridings became fatigued and suffered a fatal heart attack.

7/18/2000   **Steven Max Wilmot, Captain**   **Career, Age 47**
                      Springfield Fire Department, IL   **Multiple Organ System Failure**

Captain Wilmot was a fire investigator working on the scene of a previous structure fire. As Captain Wilmot and another fire investigator worked on the scene, Captain Wilmot caught his foot on an object and fell forward onto his chest, landing on a concrete walkway. After being helped up by the other fire investigator, Captain Wilmot said that he had fallen on his camera. The lens of the camera had created an impression on his torso. Captain Wilmot told the other fire investigator that he thought he had bruised a rib. Wilmot reported the fall to his employer and saw a doctor on the day of the fall. He was prescribed pain medication and placed on restricted duty.

Unbeknownst to him, when he fell, Captain Wilmot injured his spleen and developed a stress ulcer. The ulcer eventually perforated, releasing bowel contents into Captain Wilmot's abdomen, which caused a massive infection. Captain Wilmot became ill and was admitted to the hospital, where he died on 8/9/2000. The cause of death was due to multiple organ system failure due to peritonitis with severe hypertension, ischemic necrosis of the liver and kidneys due to blunt force trauma of the left chest wall with splenic hematomas, and a perforated stress ulcer.

The cause of the original structure fire was listed as suspicious. Four children were later arrested and charged with arson and criminal damage to property.

8/2/2000   **Richard Stark, Ambulance Captain**   **Volunteer, Age 62**
                     Thornhurst Volunteer Fire and Rescue Company, PA   **Heart Attack**

Captain Stark and members of his fire department responded to provide care for an elderly female in respiratory arrest. The removal of the patient through her house to the ambulance had been very difficult. Captain Stark climbed into the ambulance and sat in the captain's chair. At this time he experienced a heart attack. The ambulance was already en route to a rendezvous point to meet paramedics. The ambulance continued to the rendezvous point, where paramedics there treated Captain Stark and continued transport to the hospital, where he died.

8/3/2000   **Phillip Arthur "Pip" Conner, Seasonal Firefighter**   **Wildland Seasonal, Age 29**
                     National Park Service, Lake Meade National   **Helicopter Crash**
                         Recreation Area, NV

Firefighter Conner was a passenger in a Bell Ranger helicopter that was preparing to return to base for the night after helping to fight the Charlie fire. As the helicopter lifted off, it veered violently to the right and the rotor blades made contact with the ground. The helicopter came to rest back on its skids and the pilot shut the engine down. Firefighter Conner was wearing a seat belt at the time of the accident; despite this, he was severely injured and died. The other passenger and a crewmember on the ground that came to their aid were also injured. Additional information about this aircraft accident can be obtained at the National Transportation Safety Board Website under report number LAX00GA286.

8/3/2000   **Jack Elias Gazalie, Firefighter**   **Volunteer, Age 46**
                     Adamsburg & Community Volunteer Fire Department   **Trapped**

Firefighter Gazalie was killed in a fire in his home after rescuing his daughter and mother.

8/6/2000   **Bradley Scott Pierce, Firefighter/Paramedic**   **Career, Age 27**
                     Saint Charles City Fire Department, MO   **Heart Attack**

Firefighter Pierce had finished a 24-hour shift and was participating in fire department-approved physical fitness activities in the basement of his fire station. Firefighter Pierce was alone in the workout area. During the shift, Firefighter Pierce had responded to an emergency medical call and a false alarm. At some point during his workout, Firefighter Pierce suffered a heart attack. Other firefighters discovered him in seizures and provided immediate medical help, to no avail. Firefighter Pierce was a member of his department's Combat Challenge Team.

8/9/2000	**Lisa Ann Farrow, Firefighter**	**Volunteer, Age 30**
	Engelhard Fire and Rescue, NC	**Pulmonary Edema**

Firefighter Farrow had provided support at the scene of a fire that was confined to food on the stove. Firefighter Farrow had complained of the heat that day. The temperature was over 90 degrees with significant humidity. As she was returning equipment to the apparatus at the conclusion of the incident, she collapsed. EMS assistance was on scene and provided aid immediately. Firefighter Farrow was transported to the nearest hospital, 50 miles away. Shortly after her departure from the scene, Firefighter Farrow went into cardiac arrest. The cause of death was listed as acute hypoxia due to pulmonary edema. Firefighter Farrow had a history of heart problems.

8/11/2000	**James Alan Burnett, District Forester**	**Wildland Career, Age 51**
	Department of Agriculture, Forestry Services, OK	**Overrun by Wildfire**

Four times during the summer of 2000, District Forester Burnett received leave from his full-time job in Oklahoma to work as a temporary firefighter for the Federal government. He had served two assignments in Florida and one assignment in Louisiana. On 8/2/2000, he accepted another assignment as the engine boss of an Oklahoma contract engine and was eventually assigned to the "Kate's Basin" fire in Wyoming. District Forester Burnett's engine was assigned to assist local firefighters with a burnout operation. As District Forester Burnett was sizing up the fire line, a sudden wind caused the fire to "blow up." District Forester Burnett attempted to start the pump on his engine to protect his position but was unable to start the pump. District Forester Burnett attempted to reach a safety zone and attempted to deploy his fire shelter, but was unsuccessful. District Forester Burnett was wearing brush gear at the time of his injury. He died of burns.

8/12/2000	**Logan D. Fields, Assistant Chief**	**Career, Age 51**
	Hazard Fire Department, KY	**Heart Attack**

Assistant Chief Fields was on duty in the fire station. He was walking from the bunkroom to the hallway when he fell to the floor, the victim of a heart attack. He died later that day.

8/13/2000	**Lester Lee Shadrick, Captain**	**Wildland Contract, Age 53**
	ERA Aviation, Bureau of Land Management, NV	**Helicopter Crash**

Captain Shadrick was working the Twin Peaks fire Northeast of Fallon, Nevada. He was piloting a Bell 412 helicopter and was the lead chopper in a flight of two helicopters preparing to make a water drop on a fire-involved ridgeline. The helicopter was carrying a bambi bucket suspended below the aircraft. As Captain Shadrick approached the ridgeline, his aircraft made a sudden 90-degree left (descending) turn and impacted the mountainous terrain. No radio communication was received from his helicopter after to the turn and before the crash. Captain Shadrick was killed instantly.

8/13/2000	**Warren (J.C.) Smith, Private**	**Career, Age 28**
	Indianapolis Fire Department, IN	**Barotrauma**

Private Smith was participating in dive training exercises at a local quarry. Private Smith and his company were simulating the rescue of a drowning child in 70 feet of water. Private Smith had been a certified team diver for 2 years. Private Smith failed to surface with his buddy and rescue attempts were commenced immediately. CPR was started in the water and continued during a boat ride to shore. Paramedics from a local rescue squad provided aid, and Private Smith was transported by helicopter to a local hospital where he was pronounced dead. Private Smith died of barotrauma, a condition caused when a diver rises to the surface of the water too quickly and suffers internal injuries as a result of gas expansion during the ascent.

8/13/2000   **Grant F. Trick, First Assistant Chief**   **Volunteer, Age 49**
                         Canton Fire Department, PA   **Heart Attack**

First Assistant Chief Trick was in his fire station gathering firefighters to perform a controlled burn of some brush near a residence. The burn had been planned in advance and had been requested by a local resident. As he prepared for the activity, First Assistant Chief Trick suffered a heart attack. Other firefighters in the station summoned paramedics, and First Assistant Chief Trick was transported to a hospital. He died later that day.

8/14/2000   **James Robert Renfroe, Assistant Chief**   **Volunteer, Age 47**
                         Dallas County Fire & Rescue Services, TX   **Cardiac**

Assistant Chief Renfroe responded in a minipumper to a fire that involved a 240-foot long wooden railroad trestle. He worked on the scene for approximately 6 hours acting as a sector officer and pump operator. Near the conclusion of the incident, the incident commander instructed him to bring the minipumper back to the station to provide coverage for his area. Assistant Chief Renfroe got into the vehicle, started it, and then turned to the passenger and told him that he was not felling well. EMS personnel on scene were called and Assistant Chief Renfroe collapsed. Firefighters found no pulse or respiration and CPR was begun. Assistant Chief Renfroe was transported to a local hospital by air ambulance. He was pronounced dead at the hospital. The cause of death was listed as atherosclerotic cardiovascular disease. Assistant Chief Renfroe was an Equipment Supervisor for the City of Dallas Fire Department. The railroad trestle fire was believed to be accidental, caused by a passing train.

8/17/2000   **Robert Wayne Crump, Firefighter**   **Career, Age 37**
                         Denver Fire Department, CO   **Drowning**

Firefighter Crump and members of his squirt company were directing traffic away from an area that had been flooded by a very heavy rain. Firefighter Crump was wearing full structural protective clothing including a protective coat, protective trousers, and a helmet. According to the police report, 2-1/2 inches of rain had fallen in the 2 hours prior to this incident. As the firefighters were working, a woman who was attempting to cross a flooded area stalled her car in the high water and was attempting to walk to a nearby bank to make a phone call. She attempted to cross a rain-filled ditch and fell into the water. She became stuck in a pool of water that covered a culvert but was able to grab onto a pipe to prevent being drawn underwater. Unknown to anyone on the scene, the ditch led to a 64-inch concrete drainpipe that was not equipped with any type of grating. Firefighter Crump and another firefighter were summoned by the calls of citizens; both entered the water to rescue the woman. As they made their way to the woman, Firefighter Crump was immediately drawn under the water. Citizens assisted the other firefighter from the water, he returned to rescue the woman, and then turned his efforts toward attempting to locate Firefighter Crump. Approximately 5 hours later, Firefighter Crump's body was located by a police officer near an outlet of the storm water system. His cause of death was listed as drowning.

8/20/2000   **John Paul "J.P." Pritchett, Sr., Forestry Crew Chief**   **Career, Wildland, Age 56**
                         Mississippi Forestry Commission, Webster County, MS   **Overrun by Wildfire**

Forestry Crew Chief Pritchett and a forester from the wood products company that owned the land that was on fire were teamed together. The assignment for the pair was to use a tractor-plow (operated by Forestry Crew Chief Pritchett) to cut a firebreak to tie in the rear and contain the right flank of the fire. As the tractor-plow worked, the brush between the tractor line and the burned area was set ablaze by the forester using a drip torch so that future spread could be prevented.

The backfire became too intense, so the decision was made to stop the backfire part of the operation. As the forester continued to follow the tractor-plow, he encountered a bee's nest that had been plowed through by the tractor-plow. The forester, who was allergic to bee stings, made attempts to get through the area but was forced to return to the road to avoid them.

About the same time, Forestry Crew Chief Pritchett made a turn toward the fire in an attempt to locate the perimeter. Visibility was poor due to intense undergrowth and smoke. He inadvertently positioned himself in front of a finger of the fire that was making a rapid run. By the time he saw the crowning head fire rolling toward him, it was too late for a retreat. He used his dozer to create a safety zone. He laid face down in the center and covered himself with dirt in an attempt to protect himself as the fire passed over him. Forestry Crew Chief Pritchett was exposed for about 15 seconds. Forestry Crew Chief Pritchett rose from the ground, extinguished a small fire involving his tractor-plow, and drove the tractor-plow out to a point where he met some other firefighters. He sustained second- and third-degree burns to his arms, back, neck, and face.

Forestry Crew Chief Pritchett was transported to a local hospital by a local police chief and later transferred to a burn center where he was treated for his injuries. He died suddenly and unexpectedly on 11/3/2000, 2 weeks after his injury. The cause of death was listed as massive bilateral bronchial pneumonia as the result of thermal burns with hospital immobilization.

Forestry Crew Chief Pritchett was either not equipped with or failed to use a fire shelter. The county chief medical examiner's statement strongly recommended that all Mississippi Forestry Commission wildland crews be equipped with appropriate fire retardant/resistant protective clothing. The medical examiner stated that Forestry Crew Chief Pritchett would likely not have sustained his specific burn injuries had he been wearing protective equipment.

The County Line fire eventually consumed 288 acres. This fire and several others in the area were caused by arson.

8/23/2000	Michael Todd Bishop, Inmate Firefighter	Wildland Part-Time, Age 27
	Rodgie R. Braithwaite, Inmate Firefighter	Wildland Part-Time, Age 26
	Flame-N-Go Handline Unit, Utah State Prison, UT	Electrocution (Lightning Strike)

Firefighters Bishop and Braithwaite were members of a 20-person, Type 2, hand crew assigned to the North Stansbury fire, about 40 miles west of Salt Lake City. The crew had been transported by helicopter to the fire line for work. The crew was assembled for work at about 11:25 a.m. At just after noon, a squad of six firefighters, including Firefighters Bishop and Braithwaite, proceeded toward their assigned work area. A storm cell with lightning, heavy rain, and marble-sized hail moved into the area. Firefighters Bishop and Braithwaite and another firefighter took refuge under nearby trees, and the rest of the squad moved to a lower point. Lightning struck the trees where the firefighters were located.

Firefighters Bishop and Braithwaite were found in respiratory arrest. The other firefighter was injured. Emergency procedures, including CPR, were initiated, and the injured firefighters were removed to the hospital by helicopter. Despite these efforts, Firefighters Bishop and Braithwaite were later pronounced dead. The cause of death for both firefighters was listed as electrocution due to lightning.

| 8/26/2000 | Jaime Quinones, Jr., Firefighter | Career, Age 38 |
| | Waterbury Fire Department, CT | Shot |

Firefighter Quinones and members of his engine company were parked near a city park conducting a "Fill the Boot" fund drive for the Muscular Dystrophy Association. During the collection activities, Firefighter Quinones was shot several times by a gunman in a car that pulled up close to the engine apparatus. Other firefighters summoned EMS and began to treat Firefighter Quinones, who was transported to a hospital, where he was pronounced dead. The person responsible for the shooting killed two other people and himself during a 15-minute shooting spree. The perpetrator was the former husband of Firefighter Quinones' fiancé.

| 8/27/2000 | Frank Funston, Firefighter | Volunteer, Age 47 |
| | Kootenai National Forest, MT | Unknown |

Firefighter Funston was working a Montana wildfire when he became ill and was assisted from the fireline. He went to the hospital for treatment then returned to his hotel where he died 2 days later of unknown causes.

9/1/2000   Albert Leonel Voris, Jr., Lieutenant   Volunteer, Age 63
               Combine Volunteer Fire Department, TX   MVA

Lieutenant Voris was responding to the fire station as the driver and sole occupant of his personal vehicle after his department was dispatched to the report of a vehicle fire. An oncoming vehicle crossed the centerline of the roadway and struck Lieutenant Voris' vehicle head on. Firefighters at the scene of the car fire responded to a page for the vehicle accident and upon their arrival found Lieutenant Voris trapped. Lieutenant Voris was pronounced dead at the scene prior to the completion of extrication. The driver of the other vehicle received minor injuries. Lieutenant Voris was wearing a seat belt at the time of the collision. The cause of death for Lieutenant Voris was listed as multiple blunt force trauma. The cause of the car fire was listed as suspicious.

9/4/2000   Earnest Otis Ousley, Lieutenant   Career, Age 49
               Roselle Fire Department, NJ   Heart Failure

Lieutenant Ousley was assigned for the day to dispatch duties. While dispatching an alarm, Lieutenant Ousley suffered severe shortness of breath and chest pains. Firefighters treated Lieutenant Ousley, paramedics were called, and he was transported to the hospital. Approximately 6 hours after he became ill, Lieutenant Ousley died. The cause of death was listed as heart failure. Lieutenant Ousley had been hospitalized with shortness of breath a month before his death but had been released to limited duty.

9/4/2000   Daniel H. Yanklin, Lance Corporal   Career Military, Age 21
               Marine Corps Air Station Yuma, AZ   Struck by Apparatus

Lance Corporal Yanklin was performing morning checks on his assigned Airport Rescue Fire Fighting (ARFF) vehicle. As a part of the pump test, he was flowing a handline. As he was operating the handline, he was struck by another ARFF vehicle and killed. According to the commanding officer, horseplay and excessive speed were involved in the incident.

9/5/2000   Howard William Vanhoy, Assistant Chief   Volunteer, Age 67
               Austin Volunteer Fire Department, NC   Heart Attack

Assistant Chief Vanhoy was operating a pump at a live fire training exercise being conducted by his department and one other fire department. Approximately 1-1/2 hours into the exercise, Assistant Chief Vanhoy fell to the ground, the victim of an apparent heart attack. CPR was started immediately by other firefighters, and Assistant Chief Vanhoy was transported to a local hospital. He was pronounced dead shortly after his arrival. Assistant Chief Vanhoy's son Billy was the Fire Chief of the department.

9/7/2000   Michael Robert Fossett, Crew Chief   Wildland Career, Age 45
               David Timothy Newman, Pilot   Wildland Career, Age 40
               North Carolina Division of Forest Resources, NC   Helicopter Crash

Crew Chief Fossett and Pilot Newman were traveling to a public education event in their 1974 UH-1H Huey helicopter. As the aircraft entered Balsam Gap, heavy fog was encountered. The pilot radioed that he was going to land the helicopter until the fog lifted. Shortly after the radio transmission, a rotor struck a tree about 20 feet from the top of a mountain. The rotor was destroyed, and the helicopter came to rest in an inverted position and caught fire. The cause of death for Crew Chief Fossett and Pilot Newman was massive head and body trauma. Additional information related to this incident may be found in National Transportation Safety Board Accident Investigation MIA00GA264.

9/11/2000   Michael Kenneth Yahraus, Firefighter/Paramedic   Career, Age 32
                Sarasota County Fire Department, FL   Shot (Training Accident)

Firefighter/Paramedic Yahraus was participating as a SWAT/medic member of a police SWAT team. The team was practicing high-risk traffic stops. Firefighter/Paramedic Yahraus was the driver of one of the police vehicles used in the simulation and was standing on the driver's side of his car after the stop had been made. Another officer, who was playing the role of the suspect in this situation, exited his vehicle and then turned and fired a single shot at the officers who were acting in the role of police officers. The weapon used was a 38-caliber pistol loaded with blanks. The firing of the blank dislodged a lead plug that was installed in the barrel of the training weapon. The lead plug broke the windshield, ricocheted off the window post, and struck Firefighter/Paramedic Yahraus in the left eye area. Officers began to provide first aid to Firefighter/Paramedic Yahraus while EMS resources were summoned. Upon their arrival, Sarasota County paramedic firefighters provided ALS care and transported Firefighter/Paramedic Yahraus to a helicopter landing site for his trip to the hospital. He was airlifted to the hospital, where he died the next day. Firefighter/Paramedic Yahraus was in the final week of a 5-month law enforcement training program.

An investigation revealed that blank cartridges should not have been used in the training weapon. Gas expelled by the blank when it is fired and debris such as wadding that is in the blank can create pressure and force the lead plug out of the gun. The proper cartridge for use in the training weapon was a primer round.

9/17/2000	**Robert Wilson Humphrey, Firefighter**	**Volunteer, Age 62**
	Maryland Line Volunteer Fire Company, MD	**Struck by Vehicle**

Firefighter Humphrey responded to the scene of a motor vehicle accident in his personal vehicle. He parked his car on the right shoulder of the highway and began to cross the road to assist a battalion chief who had already arrived on the scene. As Firefighter Humphrey crossed, a mid-size sedan struck him. Firefighters arriving in response to the original incident assisted with the treatment of the original accident victim and Firefighter Humphrey. Firefighter Humphrey and the victim of the original accident were transported to the hospital by helicopter. Firefighter Humphrey died later that day in surgery.

9/19/2000	**George David Butler, Assistant Chief**	**Volunteer, Age 47**
	Idalou Volunteer Fire Department, TX	**Heart Attack**

Assistant Chief Butler and members of his department responded to a truck rollover that required extrication. Assistant Chief Butler operated the department's air bags and was successful in lifting the truck off the driver so that the extrication could be completed. Shortly after the truck driver had departed the scene by ambulance, Assistant Chief Butler collapsed of an apparent heart attack. Other firefighters began CPR immediately, and Assistant Chief Butler was transported to a regional hospital where he was pronounced dead 2 hours later. Assistant Chief Butler had no previous history of major illness.

9/21/2000	**Bernard D. (Pete) Scannell, Fire Police Captain**	**Volunteer, Age 70**
	Waterloo Fire Department, NY	**Heart Attack**

Fire Police Captain Scannell was driving a rescue truck to the scene of a reported car fire. As the unit responded, Fire Police Captain Scannell suffered a heart attack. The rescue truck left the roadway, jumped a curb, and came to a stop in a small flowerbed. Other firefighters immediately removed Fire Police Captain Scannell from the truck and began CPR. An ambulance arrived shortly thereafter and applied a defibrillator. After a shock was administered, a pulse was detected. Fire Police Captain Scannell was transported to the hospital where he died about 1 hour later.

9/24/2000	**Kevin Scott Harshbarger, Firefighter/Secretary**	**Volunteer, Age 36**
	Scenic Loop Volunteer Fire Department, TX	**Smoke Inhalation**

Firefighter Harshbarger and members of his department responded to a structure fire in a residence. The fire was in the attic area. Firefighters made an attempt at an interior attack but were forced from the building by extreme heat and smoke. The order was given to open the roof for ventilation. Firefighter Harshbarger and another firefighter went to the roof of the structure to cut a hole. As the hole was being cut, Firefighter Harshbarger fell through the roof into the main body of fire. Firefighter Harshbarger was not wearing an SCBA. The cause of his death was listed as smoke and soot inhalation.

| 9/27/2000 | Paul Antonio Lyndell Husband, Sr., Firefighter | Career, Age 33 |
| | Mobile Fire–Rescue Department, AL | Struck by Apparatus |

Firefighter Husband was working an overtime shift when he and members of his ladder company were dispatched to provide vehicle extrication services at a motor vehicle accident with injuries. As the ladder apparatus was leaving the station to respond to the emergency, Firefighter Husband attempted to board the apparatus. Firefighter Husband chased the apparatus on foot as it crossed two lanes of traffic and made a left turn. He had hold of a handle near the cab on the driver's side when he lost his grip, fell, and was run over by the apparatus. The members of the ladder company provided emergency medical care and additional assistance was called. Firefighter Husband was transported to the hospital by ambulance. He was pronounced dead about 1/2 hour after the incident. The cause of death was listed as multiple blunt force injuries.

| 10/1/2000 | Thomas G. Gotkowski, Captain | Volunteer, Age 55 |
| | Tinley Park Volunteer Fire Department, IL | Heart Attack |

Captain Gotkowski and his engine company were assisting the police department with the ventilation of a condominium. A resident of the home had passed away of natural causes 2 weeks previously. Captain Gotkowski assisted with the placement of a ventilation fan and assisted with the repositioning and stacking of fans. Captain Gotkowski felt ill and was sitting on the back of an engine. He walked to a nearby ambulance where it was determined that he was experiencing a heart attack. He was transported to the hospital where he died. This incident was the fifth call of the shift for Captain Gotkowski

| 10/8/2000 | Albert Roger "Bo" Rathbun, Firefighter | Volunteer, Age 69 |
| | Sundance Volunteer Fire Department, WY | Overrun by Wildfire |

Firefighter Rathbun was severely burned while cutting a firebreak with hand tools. A change of wind "blew up" a pile of smoldering debris at a wildland fire in an area that was thought to be safe. Firefighter Rathbun attempted to outrun the advancing flames, but he fell and was severely burned. Firefighter Rathbun was transported to a burn unit at a hospital in Greeley, CO, with third-degree burns over 40 percent of his body and second-degree burns over 10 percent. Firefighter Rathbun suffered a stroke while in the hospital and died on 11/8/2000.

In an interview that he gave to a local newspaper the day before the fire, Firefighter Rathbun said that he had fought his last fire and that he was retiring from his department. The day after the interview, Firefighter Rathbun and his son were working on their ranch when they saw smoke, they responded and helped to control the fire.

| 10/10/2000 | Richard J. LeClair, Captain | Career, Age 53 |
| | Federal Fire Department San Diego, CA | Internal Hemorrhage |

Captain LeClair had returned to duty after a 6-month battle with cancer. During his fourth workday since his return, Captain LeClair became ill with flu-ike symptoms on duty and was transported to a hospital by helicopter. Captain LeClair died the next day as a result of internal hemorrhaging.

| 10/13/2000 | David C. Fitzgerald, Firefighter | Career, Age 63 |
| | Somerville Fire Department, MA | Heart Attack |

Firefighter Fitzgerald and his ladder company responded to assist with treatment and cleanup at a collision involving a tractor/trailer and a recycling truck. Firefighter Fitzgerald assisted with patient treatment and packaging and then assisted other firefighters as 55 bags of absorbent were distributed over a fuel spill. The incident lasted for over 2 hours. At the scene of the collision, Firefighter Fitzgerald complained of shoulder pain, but dismissed it as a strain. He collapsed at the fire station after another emergency incident due to a heart attack. Other firefighters attempted to revive him. Firefighter Fitzgerald was rushed to the hospital by ambulance but was pronounced dead on arrival.

10/16/2000 Kenneth T. Miller, Captain   Volunteer, Age 65
              Cape Charles Volunteer Fire Company, VA   Heart Attack

Captain Miller was the backup person on a 2-1/2-inch line that was being operated on a three-story wood-frame residence. Captain Miller collapsed and medical care was immediately initiated by the firefighter that had been on the nozzle. Captain Miller was treated by EMS personnel on the scene and transported to the hospital. He was later pronounced dead at the hospital, the victim of a heart attack. The fire was caused by arson.

10/26/2000 James Reavis, Captain   Volunteer, Age 69
              North Stone Northeast Barry County Fire Protection   Heart Attack
                District, MO

Captain Reavis was the first firefighter to arrive at the scene of a residential fire. As fire apparatus began to arrive, Captain Reavis assisted with stretching lines and setting up equipment. Once things were underway, Captain Reavis drove his personal vehicle to the fire station to retrieve the department's tanker (tender). The station was a short distance from the fire. Captain Reavis was stricken with a heart attack and ran into a parked pickup truck that belonged to a firefighter working on the scene. Several firefighters were diverted from the fire to provide aid to Captain Reavis. Despite their efforts, Captain Reavis was pronounced dead at the scene.

10/31/2000 Robert M. Samanas, Firefighter/Paramedic   Part-Time, Age 52
              Rural/Metro Fire Department, Bethlehem Steel, IN   Cardiac

Firefighter/Paramedic Samanas had completed his yearly physical agility test and stopped to take a break. About 40 minutes after completing the test, Firefighter/Paramedic Samanas returned to assist other firefighters with the test. At this point, he became short of breath. He was placed on oxygen, started feeling better, and then began to experience chest pain. ALS cardiac procedures were started; however, Firefighter/Paramedic Samanas collapsed before a monitor defibrillator could be attached. Firefighter/Paramedic Samanas was transported to a local hospital where he later died. No autopsy was performed.

11/2/2000 Jared Conner McCormick, Firefighter   Volunteer, Age 19
             Bono Fire Protection District, AR   Struck by Vehicle

Firefighter McCormick attended the weekly meeting and work night at his fire department. It was determined that a piece of apparatus was in need of fuel. Two firefighters, including Firefighter McCormick as the passenger, mounted the truck and headed for a fuel station. On the way, the apparatus stalled and could not be restarted. Firefighter McCormick radioed the fire station, told them about their vehicle trouble, and requested assistance. Other firefighters brought another fire truck and the chief's pickup to the location of the broken truck. After the truck was removed from the roadway, an attempt was made to boost or jump-start the fire truck using the chief's vehicle. When this failed, it was decided that the broken truck would be towed back to the station by the other fire truck. The chains needed for the job were aboard the other fire truck. At this time, the broken truck and the chief's vehicle were off the road and the fire truck that was to tow the broken truck back to the station was parked across the street due to construction in the area. The area was dark, and the four-way flashers on the chief's vehicle and the working fire truck were in operation.

Firefighter McCormick began to cross the road to retrieve the chains. Firefighter McCormick signaled to an approaching minivan to stop. As his attention was focused on the minivan, a tractor/trailer that approached from the other direction struck him. Firefighter McCormick was thrown into the path of the minivan and was struck a second time. Firefighters on the scene rushed to Firefighter McCormick's aid and he was transported to the hospital, where he was pronounced dead. The cause of death was listed as massive blunt trauma.

11/2/2000 Gail Lynne VanAuken, Firefighter   Volunteer, Age 41
             Overisel Township Fire Department, MI   Apparatus Rollover

Firefighter VanAuken was the passenger in a tanker (tender) responding to a mutual-aid structure fire involving a turkey farm. Firefighter VanAuken's husband was driving the 2,000-gallon tanker with lights and siren in operation. As the apparatus approached an intersection, a pickup truck approaching the intersection from the other side street appeared to be yielding the right of way to the tanker. The tanker slowed before going through the stop sign. As the tanker proceeded through the intersection, it was struck by the pickup at the left rear axle. The tanker rolled over, the water tank separated from the chassis, and both firefighters were trapped in the cab.

Firefighters from other departments responding to the fire came upon the accident scene and provided aid. Both firefighters were extricated from the cab and transported to the hospital by ambulance. The extrication took about 30 minutes. The injuries to the other firefighter and the driver of the pickup were not life threatening. Firefighter VanAuken received crushing blunt force chest injuries; her cause of death was listed as mechanical and positional asphyxiation.

| 11/9/2000 | James G. Hill, Sr., Firefighter/Safety Officer | Volunteer, Age 67 |
| | Daingerfield Volunteer Fire Department, TX | Heart Attack |

Firefighter/Safety Officer Hill responded with other members of his department to a mutual-aid structure fire involving a mobile home. Firefighter/Safety Officer Hill assisted with support duties on the fireground and moved inside of the mobile home to assist other firefighters that were performing overhaul. There was a very light smoke condition inside the mobile home. Firefighter/Safety Officer Hill became short of breath and on-scene paramedics began treatment. While en route to the hospital by ambulance, Firefighter/Safety Officer Hill suffered a heart attack. He was revived but suffered another heart attack in the hospital and died.

| 11/15/2000 | Kenneth W. Kerr, Firefighter | Career, Age 44 |
| | New York City Fire Department, NY | Heart Attack |

Firefighter Kerr and members of his engine company had just returned from fighting a stubborn fire in an elevator cab in a six-story building. At the scene, Firefighter Kerr told other firefighters that he did not feel well but refused medical aid. When his company returned to quarters, Firefighter Kerr spent some time with other firefighters in the kitchen and then headed for the shower. He was found collapsed in the shower by other firefighters. Medical treatment was administered immediately by other firefighters, but Firefighter Kerr died.

| 11/16/2000 | Kyle Allen Hendrick, Firefighter | Volunteer, Age 19 |
| | Gott Volunteer Fire Department, KY | Apparatus Rollover |

Firefighter Hendrick was the driver of a 1,500-gallon fire department tanker (tender) participating in a water shuttle drill. The passenger in the truck was a 17-year-old trainee. Neither Firefighter Hendrick nor the trainee was wearing a seat belt. As the tanker traveled down the road, the vehicle's right wheels dropped off the roadway. Firefighter Hendrick overcorrected to the left and came back on the road, riding the centerline. He corrected again and went off the roadway on the right-hand side. The tank separated from the vehicle, and the cab came to rest on its top. Firefighter Hendrick was partially ejected from the vehicle. The trainee was fully ejected from the vehicle. Firefighter Hendrick was removed from the vehicle and transported to the hospital by ambulance. He was pronounced dead at the hospital approximately 1 hour after the collision. The trainee was severely injured.

11/16/2000	Phillip Dewey Smith, Driver/Operator Engineer	Career, Age 49
	Department of Defense Fire Department,	Heart Attack
	Fort McPherson and Fort Gillem Fire and Emergency	
	Services, GA	

Driver/Operator Engineer Smith was a participant in wildland firefighting training. He was the crew boss and helped his crew dig a firebreak as a part of the exercise. As the work was completed, Driver/Operator Engineer Smith fell to the ground and went into seizures. Firefighters that had been involved in the exercise provided immediate care and an ambulance was summoned. Driver/Operator Engineer Smith was transported to the hospital where he died of a heart attack.

11/17/2000   **Thomas J. Hazaz, Fire Police Lieutenant**   Volunteer, Age 69
Tunkhannock Township Volunteer Fire Company, PA   Heart Attack

Fire Police Lieutenant Hazaz responded to the scene of a motor vehicle accident with members of his department. When he arrived on the scene in his personal vehicle, Fire Police Lieutenant Hazaz received orders from the fire chief by radio. As he passed the scene en route to his assignment, he waived the fire chief over to his pickup. The chief opened the door of the pickup and repeated his orders. Fire Police Lieutenant Hazaz waved to acknowledge the order and placed his hands on the wheel. The chief closed the pickup's door and noted that the vehicle did not move. The chief opened the door and discovered that Fire Police Lieutenant Hazaz was suffering a heart attack. Firefighters removed Hazaz from his pickup, CPR was started, and an ambulance was called. The ambulance that was on scene for the initial accident had already departed for the hospital. Despite efforts on the scene and on the way to the hospital, he was pronounced dead at the hospital. The cause of death was listed as atherosclerotic cardiovascular disease.

11/25/2000   **Marvin Maurice Bartholemew, Professional**   Career, Age 30
  Firefighter II   Smoke Inhalation
Pensacola Fire Department, FL

Firefighter Bartholemew responded as a member of an engine company to a report of a residential fire. Upon arrival on the scene, the first company officer reported a working fire with approximately 50 percent of the building involved. Firefighter Bartholemew was assigned to join the crew of a rescue and perform a search of the structure. A handline was stretched by the search crew and carried into the structure. Five to 10 minutes after arrival, the company officer from the rescue realized that fire was spreading behind them. He ordered his crew to abandon their efforts and leave the house. All three firefighters headed for the exit as the flashover occurred. The company officer and the firefighter from the rescue emerged from the structure; both were burned. Firefighter Bartholemew was not with them. The company officer reported Firefighter Bartholemew missing. At least four searches were completed before Firefighter Bartholemew was located—approximately 1 hour after the flashover. He had apparently become disoriented and ended up in the kitchen at the back of the house. The cause of death was listed as asphyxia due to smoke inhalation. The carboxyhemoglobin level in Firefighter Bartholemew's blood was 69.5 percent. The fire was caused when a pan caught fire on top of the stove and extended. The occupants of the house had evacuated prior to the arrival of the fire department.

11/26/2000   **Daniel I. King, Firefighter**   Volunteer, Age 21
Cliffside Park Fire Department, NJ   MVA

Firefighter King was responding in his personal vehicle to an automatic fire alarm. He was not displaying emergency or courtesy lights, but he was flashing his headlights and honking his horn. As he responded, a vehicle emerged from a side street on his right. Firefighter King swerved into the oncoming lane to avoid the collision, his vehicle began to fishtail, and he hit a transit bus head on. Firefighters responded to the scene and extricated Firefighter King from his vehicle; he died later that day. The cause of death was listed as internal trauma. Firefighter King was wearing a seat belt.

11/29/2000   **Elwood Queen, Firefighter**   Volunteer, Age 67
Irvona Volunteer Fire Company, PA   Apparatus Rollover

Firefighter Queen was the driver of a fire department ambulance that was transporting a cardiac arrest patient to the hospital. As the ambulance was en route to the hospital, Firefighter Queen experienced a heart attack. The ambulance left the road, hit a utility pole, rolled 2-1/2 times, and ended up in its roof. A paramedic and two EMTs riding in the ambulance received minor injuries and provided treatment for Firefighter Queen. They were able to revive him on the scene, but Firefighter Queen died the next day. The patient that was being transported expired on the scene. The ambulance involved in the accident was 2 months old and was destroyed.

12/1/2000   **George H. Cardozo, Firefighter**   **Volunteer, Age 80**
                      Westport Volunteer Fire Department, CT   **Heart Attack**

Firefighter Cardozo worked on the scene of a residential structure fire in his role as fire department photographer. At the scene of the incident, Firefighter Cardozo complained of indigestion but refused help from EMS personnel at the scene. He returned home at the conclusion of the incident and suffered a heart attack after midnight. A police officer was the first to reach Firefighter Cardozo's home and applied an AED. Firefighters provided CPR and assisted EMS personnel. Firefighter Cardozo was transported to a local hospital where he was pronounced dead. The fire was caused by arson. December 1st was Firefighter Cardozo's 50th wedding anniversary. He responded to the fire after the celebration.

12/11/2000   **Edward A. Russ, Firefighter**   **Volunteer, Age 39**
                      Bethel Volunteer Fire Department, VT   **Struck by Vehicle**

Firefighter Russ was on his way to work when he stopped to assist the occupant of a car that had spun out of control and hit a guardrail. Firefighter Russ found that the occupant of the vehicle was not injured severely, and he turned his attention to directing traffic around the car to avoid subsequent collisions. After only a few minutes on the scene, Firefighter Russ was struck by a pickup truck that was traveling at a speed estimated at 70 miles per hour. He was killed instantly. A state highway truck was also on the scene. The driver had just begun preparations to move the truck to the other side of the road to protect the site of the original collision. The driver of the vehicle involved in the original collision was intoxicated and was later charged with driving while intoxicated.

12/17/2000   **Charles E. H. Lauber, Jr., Commissioner**   **Volunteer, Age 55**
                      Smithtown Fire Department, NY   **Trauma**

Commissioner Lauber and other members of his department had just completed the department's annual Christmas parade. Commissioner Lauber was on top of a fire truck attempting to reset the motor of a garage door opener that had malfunctioned. He fell off the top of the truck and suffered a severe head injury. Commissioner Lauber was transported to the hospital where he died on 12/24/2000.

12/17/2000   **Keith P. Purcell, Firefighter**   **Volunteer, Age 47**
                      Southold Fire Department, NY   **Heart Attack**

Firefighter Purcell and members of his department responded to a report of a structural fire. As Firefighter Purcell advanced a hoseline toward a fully involved detached two-car garage, he collapsed. Other firefighters came to his aid immediately and CPR was started. Members of the Southold Fire Department provided ALS care and transport to the hospital. Firefighter Purcell was pronounced dead approximately 1 hour later at a local hospital. The cause of death was a heart attack. Firefighter Purcell had been diagnosed with leukemia several years prior to his death.

12/18/2000   **Ronald Haner, Deputy Chief**   **Career, Age 61**
                      Portage Fire Department, MI   **Smoke Inhalation**

After warning his wife, Deputy Chief Haner was overcome by smoke and fumes while trying to escape from a structure fire at his residence.

12/23/2000   **David A. Anderson, Firefighter**   **Career, Age 43**
                      Manchester Fire Department, NH   **Heart Attack**

Firefighter Anderson responded with his engine company to a structure fire involving a three-family residence. Upon arrival, firefighters found a working fire with reports of people trapped inside. Firefighter Anderson assisted with fire control and search-and-rescue functions. Two unconscious boys were located and removed from the fire building by firefighters. After 20 minutes inside the structure, Firefighter Anderson came outside, sat on the rear step of an engine, stood up, and collapsed. Firefighters provided assistance immediately and Firefighter Anderson was transported by ambulance to the hospital, where he died. The cause of death for Firefighter

Anderson was listed as a heart attack. Two boys, ages 14 and 17, were also killed in the fire, which was caused by an overloaded electrical extension cord. The 17-year-old boy had reentered the house in an attempt to save his younger brother.

12/23/2000   Scott P. Gillen, Lieutenant   Career, Age 37
**   Chicago Fire Department, IL   Struck by Vehicle**

Lieutenant Gillen and members of his truck company were dispatched to the site of a motor vehicle collision on an expressway to provide a traffic shield with their apparatus and to assist ambulance personnel. Two state police cars were positioned behind the ladder in a further attempt to divert traffic. As the incident was being concluded, Lieutenant Gillen walked around the truck to make sure that everything was ready to go. As Lieutenant Gillen walked, a passenger car ran over a line of flares in an attempt to slip by traffic. The car then struck a tractor/trailer truck, spun, and pinned Lieutenant Gillen between the car and the ladder truck. Lieutenant Gillen was treated at the scene and then airlifted to the hospital. His legs were crushed in the collision, and he had lost a lot of blood. He died 10 hours later.

The driver of the car that struck Lieutenant Gillen was under the influence of alcohol and was driving on a suspended driver's license. He was later charged with reckless homicide. There were no injuries in the original collision.

Alphabetized Firefighter Fatalities and Date of Incident

Acey, Vencent, 1/28/1994
Adams, Robert L., Sr., 5/17/1990
Adams, John R., 10/29/1991
Adams, Norman, 4/81996
Adams, Stanley, 4/24/1996
Adkins, Raymond, 4/30/1993
Ainsworth, James "Frank", 10/28/1995
Alexander, Mathe A., 5/7/1990
Alfred, Joseph J., 2/16/1991
Allen, Timothy D., 6/29/1998
Allgood, Harold B., 11/15/1993
Almond, Norman Neal, 11/24/1998
Altic, James William, 1/17/2000
Altieri, Ronald C., 1/19/1991
Alves, Lionel, 8/3/1992
Anderson, Vidar D., 2/19/1990
Anderson, Kaye F., 4/19/1990
Anderson, Lewis Edward "Rawhide", 9/27/1999
Anderson, David A., 12/23/2000
Archer, Thomas James, Jr., 4/9/1998
Armstrong, Gary W., 11/15/1993
Arnone, Craig, 12/8/1996
Arthur, Eddie, 6/1/1991
Augustin, Walter, 11/7/1995
Ayers, Wendell, 2/14/1995
Ayers, Carl L., 1/7/1997
Babka, William W., 4/20/1997
Bachman, Sandra J., 6/26/1990
Bacon, Richard Clarence, 8/5/1999
Baker, Loren E., 4/15/1993
Baker, Francis J., 10/12/1993
Baltic, Peter, 6/17/1990
Bankert, Gary Lynn, Sr., 1/15/2000
Barnes, Robert D., 9/30/1992
Barrera, David, 6/7/1995
Barter, David, 6/18/1994
Bartholemew, Marvin Maurice, 11/25/2000
Batten, Lisa, 2/13/1995
Bauerlien, Eugene, 10/19/1996
Baughn, Michael R., 4/11/2000
Bayer, Kenneth E., 9/5/1997
Beadle, Charles H., 11/20/1993
Beck, Richard, 7/6/1993
Beck, Kathi, 7/6/1994
Belcher, Randolph F., 6/21/1991
Bennett, James A., 1/21/1992
Bennett, Paula, 3/16/1998
Benton, Leon L., 12/27/1990
Berggren, Corey, 8/24/1995
Berry, Scott M., 12/17/1997
Bethune, William Malcolm, 10/5/1999
Bianconi, Thomas N., 5/17/1990
Bibbee, Robert, 6/29/1996
Bice, Martha Ann, 10/18/1996
Bickett, Tami, 7/6/1994
Biedron, Fred P., 12/16/1991
Birchmore, Floyd, 8/8/1996
Bishop, Michael Todd, 8/23/2000

Bitting, Jason L., 12/22/1999
Bjorkland, Paul K., 7/4/1992
Black, Matthew Eric, 6/23/1999
Blackmon, Eugene Williardk, Jr., 5/19/1998
Blackmore, James, 6/5/1998
Blakemore, William M., 3/8/2000
Blankenship, Arnold, III, 4/30/2000
Blanusa, George, 3/10/1993
Blecha, Scott, 7/6/1994
Blizzard, Robby Dean, 11/6/1998
Bohan, James F., 12/18/1998
Bonnar, William E., Sr., 2/25/1998
Bookout, Roger DeWayne, 11/18/1998
Boomer, Robert, 7/12/1994
Boothe, Joseph Jay, 4/2/1994
Bopp, Christopher M., 12/18/1998
Borwegan, Peter "Butch", 6/6/1995
Boster, Jonathan C., 4/8/1996
Boswell, Jospeh A., 12/26/1992
Boudoin, Keith, 10/13/1996
Bow, Joseph R., 10/21/1991
Boyce, Samual Isaac, 3/15/1993
Boyert, Anthony L., 4/21/1990
Boyle, Charles H., 6/9/1991
Bradner, William S., III, 11/5/1997
Brain, Frederick L., 3/18/2000
Braithwaite, Rodgie R., 8/23/2000
Brannon, Robert Cowey, Jr., 4/26/2000
Brashears, Thomas D., 6/19/1990
Brekrus, Richard V., 2/20/1990
Brentzel, John, Jr., 12/23/1993
Bricker, Walter L., 8/12/1996
Bricker, Jessie F., Jr., 5/3/1997
Bridges, William, 4/11/1994
Brinkley, Levi, 7/6/1994
Brinkley, David M., 9/21/1998
Broadhead, Ralph M., 8/26/1990
Brooks, Jerry, 8/8/1991
Brooks, Larry H., 12/23/1991
Brooks, Thomas, 2/14/1995
Brotherton, Paul Arthur, 12/3/1999
Broussard, Henri Fred, 11/18/1999
Brown, Mary Jo, 11/20/1994
Brown, James, 1/5/1995
Brown, Craig Daniel, 11/24/1998
Browning, Robert, 7/6/1994
Bruecher, Bert Andrew, 11/14/1999
Bryant, Steven E., 6/22/1991
Bryant, Raymond L., 7/28/1991
Bryant, Jimmy, 2/25/1995
Bryant, John, 11/9/1996
Bryant, Kendall O., 3/31/2000
Buc, Robert, 8/13/1994
Buchholtz, Evan, 12/26/1994
Buckert, Walter Douglas, 10/2/1997
Buff, Thomas, Jr., 11/11/1995
Buhler, Robert W., 3/6/2000
Bullard, Anthony, 6/29/1994

Burger, Dwight, 12/6/1994
Burkhalter, Dale, 1/26/1996
Burley, Clyde A., 11/21/1991
Burnett, Brian K., 10/28/1999
Burnett, James Alan, 8/11/2000
Buroker, Dennis L., 6/2/1998
Burton, George A. "Bo", 6/4//2000
Butchee, Charles, 3/5/1994
Butcher, Marc, 6/13/1994
Butler, Frank L., 3/26/1991
Butler, Michael A., 3/23/1998
Butler, George David, 9/19/2000
Buttram, Bill, 7/28/1995
Byers, John M., 8/25/1992
Calhoun, Anthony Vo, 10/12/1992
Campbell, Herbert B., 3/2/1992
Campbell, James Shannon, 8/31/1992
Cannon, Brian Allen, 1/16/98
Canonico, Michael, 11/25/1995
Capps, Jackson, 10/24/1996
Cardozo, George H., 12/1/2000
Carletti, Stephen D., 2/5/1998
Carlson, Ronald, 4/7/1994
Carpenter, David E., 9/8/1997
Carr, Marcus, 1/3/1994
Carrasco, Brian, 8/31/1998
Carter, William F., Sr., 11/30/1990
Carter, John M., 10/24/1997
Carter, Gregory Scott, 1/21/1998
Carugno, Anthony J., 11/1/1992
Casboni, Matthew P., 7/23/1998
Cash, Bedford, 2/26/1994
Cashman, Kenneth C., 9/13/1999
Casiano, Eric Noel, 5/3/1999
Castillo, Victor Clement, 5/7/1998
Castro, David, 8/8/1994
Cavalieri, Joseph P., 12/18/1998
Caywood, Paul E., Sr., 9/23/1990
Certain, James, 9/22/1994
Chacon, Joseph L., 6/26/1990
Chambers, William, 11/12/1996
Champney, Gordon A., 3/1/1992
Chapin, Tony B., 9/24/1998
Chappell, James A., 2/24/1991
Charmello, Nick, 1/28/1994
Chesney, Charles Brant, 12/27/1996
Chestnut, James E., Jr., 5/1/1990
Chestnut, Burton Frank, 2/18/1999
Chisholm, Robert D., 8/31/1997
Chlian, Jerome H., Jr., 7/25/1997
Christian, Loren N., 12/24/1990
Cielicki, Michael J., 12/20/1991
Ciliberto, George, 1/22/1994
Cirrito, Phillip P., 6/9/1999
Clancy, John, 12/31/1995
Clark, Mark, 4/23/1996
Clark, James Everett, III, 8/5/1999
Clawson, Corey R., 8/28/1992
Clinch, Kenneth F., 7/30/1999
Coates, Allan F., 7/14/1993
Cochran, John R., 11/29/1991
Cockrell, Gary, 6/22/1995
Cohen, Rudolf, 6/2/1999

Cole, Robert A., 12/19/1991
Collins, Bennie B., 11/11/1990
Collins, Donald, 4/19/1996
Collins, Brian William, 2/15/99
Colona, Steven, 12/27/1994
Colton, Todd D., 9/6/1990
Concannon, Thomas J., 12/18/1998
Conklin, Edwin R., 6/29/1993
Conner, Phillip Arthur "Pip", 8/3/2000
Conroy, Patricia, 2/14/1995
Contreras, Alex S., 6/26/1990
Cooper, Paul Eugene, 2/11/2000
Coppin, Jerry Wayne, 3/7/2000
Copple, Gene K., 11/18/1990
Cormican, Bruce, 8/21/1995
Cothran, James J., 12/17/1992
Cottrell, Donald C., 3/8/1993
Coulter, Leonard, 8/21/1996
Covis, Anthony, 6/4/1994
Cowgill, Brooks, 4/4/1991
Coyne, Thomas E., Jr., 12/18/1991
Craft, Willie, 4/24/1996
Crago, Roy Kenneth, 12/7/1999
Crane, George, Jr., 7/11/1996
Cribley, Kinnison F., 12/13/1991
Cropper, Leroy, 4/24/1995
Crowe, Bobby, 3/13/1995
Crown, Peter, 7/21/1995
Crump, Robert Wayne, 8/17/2000
Crutchfield, Robert T., III, 1/28/1990
Cupp, Michael Eugene "Cuppie", Sr., 8/5/1999
Cussen, Greg, 5/5/1995
Cutter, Clayton M., 2/19/1990
Dame, James E., 3/7/1991
Daughenbaugh, Donald I., 3/24/1991
Dauzat, Cilton Jay, 8/5/1999
Davis, Robbie, 12/10/1990
Davis, George A., 11/18/1997
Davis, Rickey Levi, 4/20/2000
Dawson, Timmy Roger, 8/31/1999
De Leon, Juan Gilberto, 1/17/2000
Dear, Edgar, 12/21/1990
Dearing, Dennis, Jr., 2/27/1994
Deer, Ronald, 5/24/1995
DeFlumere, Albert, 10/26/1996
DeLane, Michael, 10/29/1994
Delvecchio, Joseph D., Sr., 8/25/1993
Denney, James L., 6/26/1990
Denny, Arthur R., II, 9/28/1991
Derryberry, Clark, 10/12/1996
Deshazor, Vernon D., 1/28/1990
Detty, Earl, Sr., 9/3/1994
Devine, Richard F., 7/29/1999
Dibbles, Jerald, 1/22/1996
Dillon, Brian T., 1/28/1991
Doherty, Martin, 9/21/1996
Donahue, Jerry David, 6/27/1998
Dorr, Thomas, 1/7/1996
Dorsey, Richard, 9/4/1996
Dougherty, Edward P., Jr., 4/8/1990
Dougherty, Patrick J., 4/7/1993
Drennan, Robert J., 12/11/1990
Drennan, John, 3/29/1994

Drews, Karl J., 4/19/1990
Drobitsch, Michael F., 6/20/1997
Drury, Craig, 8/8/1994
Du Chateau, John A., 10/22/1991
Ducheck, Alan W., 2/28/1999
Dudley, Earle V., III, 9/20/1992
Dugan, John G., 7/7/1991
Dunbar, Doug, 7/6/1994
Dunham, Russell E., 7/16/1991
Dunham, James Melvin, 11/18/1999
Dunkerley, David M., 6/20/1993
Dunn, Thomas, 1/4/1994
Dupee, Joseph C., 3/8/1998
Durflinger, Vinton, 3/8/1996
Duvall, Robert, 4/7/1996
Dyer, Wilbert L., 11/8/1992
Dzioba, Heather L., 9/1/1992
Eager, Victor E., 10/16/1992
Ebert, Arthur R., 1/3/1997
Elliott, Harold "Ray", 10/15/1997
Ellis, James E., 6/26/1990
Ellis, Gwyn L., 10/20/1992
Ellis, James A., 12/21/1996
Ely, Robert J., 9/25/1990
Emanuelson, David G., 12/20/1991
Emmrich, Raymond, 12/29/1996
English, Robert, 2/5/1994
Enslow, Kenneth E., 8/13/1990
Ernst, Walter A., 5/27/1998
Ertle, Gerald T., 6/7/1997
Estavillo, Joseph J., 8/3/1997
Evans, Roger, 11/23/1994
Ezernack, Paul Franklin, Jr., 12/15/1999
Fain, Cecil A., 1/2/1993
Fairweather, Frederick, 9/23/1995
Fairweather, William H., 11/15/1997
Falconer, Lou, 3/15/1991
Farrell, Joseph D., 3/19/1991
Farrow, Lisa Ann, 8/9/2000
Favinger Sr., William R., Sr., 1/5/1996
Ferrante, Theodore A., Jr., 12/20/1999
Ferrera, Victor, 6/27/1990
Fields, Logan D., 8/12/2000
Fierson, Charles E., 9/5/1992
Fisher, John, 9/28/1995
Fitzgerald, David C., 10/13/2000
Fitzpatrick, June, 8/1/1995
Floyd, Tracy D., 5/3/1997
Flyntz, Walter J., 3/17/1999
Fogel, Wayne M., 2/4/1997
Folds, John W., 1/26/1990
Fossett, Michael Robert, 9/7/2000
Foster, Robert H., Sr., 10/29/1991
Fowler, Robert E., 2/22/97
Fowler, Vincent, 6/3/1999
Frank, William, 5/30/1996
Frank, Charles Peter, III, 11/8/1998
Franklin, Arthur Bruce, 2/28/1999
Franks, Walter, 1/28/1994
Frantz, Richard A., 12/20/1991
Freeman, Edward, 11/12/1994
Frizzell, Henry, 1/24/1995
Fullbright, Gus, 9/13/1994

Fuller, Ronnie, 1/1/1994
Funston, Frank, 8/27/2000
Garber, Fred R., 7/10/1990
Garcia, Ernestine, 2/2/1995
Garis, Harry B., 1/2/1993
Garlinghouse, Lyle, 7/20/1995
Garneau, Christopher, 8/16/1995
Garnett, Jackie Mac, 11/20/1999
Gass, Walter Harvey, 1/27/2000
Gates, Jesse, 6/3/1997
Gazalie, Jack Elias, 8/3/2000
Geary, Stoy, 1/14/1997
Geiger, James D., 2/19/2000
Gelenius, Elwood M., 2/5/1992
Gessler, Stephen E., 12/12/1998
Gilbert, Frank, Jr., 10/26/1996
Gillen, Scott P., 12/23/2000
Giradot, Steve L., 10/12/1990
Glasgow, Ralph F., 9/30/1990
Gleason, Richard E., Sr., 7/9/1993
Glenn, Marcel, 1/18/1996
Godsil, Toni J., 8/6/1990
Goessling, John, 4/23/1996
Goff, Timothy M., 5/24/1997
Golden, Bryan J., 2/6/1997
Good, Robert Douglas, 6/5/1997
Good, David John, 3/9/1998
Goode, James, Jr., 6/19/1990
Goodwin, Tulon Lee, 7/6/1998
Gosey, Carson L., Sr., 10/24/1998
Gotkowski, Thomas G., 10/1/2000
Gouckenour, Jason A., 1/9/1999
Gray, Barry M., 8/24/1990
Gray, John, 8/25/1996
Griffith, James Ted, 4/7/2000
Grimes, William P., 3/30/1990
Gross, Tommy T., III, 3/21/1997
Grosse, Jack, 10/24/1996
Grounds, William, 1/31/1993
Guerrero, Lupe, 7/27/1992
Guilmette, Ronald A., 12/11/1997
Gushiken, Steven, 2/1/1996
Gutierrez, Sean, 7/12/1994
Guyer, George, 7/28/1996
Hagen, Terri, 7/6/1994
Haggadone, Robert, 1/19/1996
Haislopp, Paul, 3/23/1999
Hale, Timothy, 2/11/1994
Halcoy, Laura, 12/21/1996
Hamilton, Barvon Coy, 8/1/1998
Hamler, Robert, 5/1/1996
Haner, Ronald, 12/18/2000
Hansen, Wilbert F., 11/11/1993
Hansen, Herloff "Ted", III, 4/13/1995
Harbaugh, Calvin, Sr., 8/18/1998
Hargreaves, John F., 8/22/1993
Harmon, Monte Jason, 6/23/1996
Harness, David, 11/19/1995
Harris, Larry R., Jr., 8/5/1993
Harris, Clifford, 6/28/1994
Harris, James, 9/1/1994
Harrison, Rufus J., 8/21/1990
Harrison, Eldon W., 9/9/1992

Harshbarger, Kevin Scott, 9/24/2000
Hart, Nicholas W., 4/6/1990
Hartley, Richard A., 1/10/1993
Hartwick, David Dwayne, 7/27/1999
Harvey, James, 8/13/1994
Hatcher, Mathew, 4/19/1996
Hathaway, H. Robert, 1/8/1997
Hauber, Russett "Rusty" S., 3/15/1997
Hauk, Brian T., 12/23/1997
Haungs, Edwin J., Sr., 6/16/1997
Havens, Richard A., 2/20/1990
Haviar, John, 11/9/1995
Hazaz, Thomas J., 11/17/2000
Hazuka, Raymond, 9/16/1991
Heckman, Arthur J., 8/9/99
Hedrick, Kenneth M., 1/12/1992
Heide, Robert S., Sr., 6/20/1992
Heimann, Frederick J., Jr., 6/13/1990
Heinze, Richard Anthony, 6/4/1999
Heirtzler, Allen Lawrence, 9/4/1998
Heller, Richard C., Jr., 4/4/1992
Henderson, Troy V., 3/19/1993
Hendrick, Kyle Allen, 11/16/2000
Hendricks, Leslie, 8/24/1996
Hennan, Clifford L., 3/4/1992
Henry, Robert, 1/25/1991
Henry, Dewey, 9/11/1994
Herington, Ken M., 6/20/1990
Herman, Christopher D., 8/20/1993
Hernandez, Juan Manuel, Jr., 9/4/1998
Hershey, Richard L., 5/1/1990
Hester, Harold, 1/10/1997
Hicks, Thomas E., 12/16/1990
Hill, Alston F., Sr., 3/28/1991
Hill, James D., 12/26/1992
Hill, Brian L., 9/6/1993
Hill, James G., Sr., 11/9/2000
Hinson, Mark R., Sr., 12/15/1993
Hinson, James Greg, 8/27/1995
Hirth, Walter John, Jr., 4/20/1997
Hiser, Floyd Dean, Sr., 7/6/1997
Hitchcock, Robert L., 3/13/1990
Hoad, Rex, 6/19/1996
Hobson, Lawrence, 5/12/1997
Hocking, Bradley, Sr., 6/3/1995
Hodges, Paul, 7/27/1994
Hoeffner, Jake M., 5/26/1998
Hoffer, Lionel, 12/24/1994
Hogan, Richard, 6/10/1995
Holcombe, David E., 2/24/1991
Hollis, Alfornia, 6/9/1991
Hollowniczky, George, 1/17/1991
Holmes, Neil A., 9/28/1998
Holmgreen, Ronald, 6/14/1994
Holsapple, Earl, 4/26/1997
Holtby, Bonnie, 7/6/1994
Hone, Jack Lee, 11/12/1994
Honstain, Bruce, 9/10/1996
Hood, Terri LeAnn, 9/15/1999
Hoots, Ronnie E., 10/19/1990
Hoover, Michael D., 10/14/1992
Hopey, George, Jr., 11/26/1997
Hopler, Willard, 1/7/1996

Hoppenjans, Ralph L., 12/22/1990
Horvath, Thomas I., 6/16/1992
Houghton, Roger A., 3/10/1990
Howe, James E., 1/9/1991
Howe, Henry W., 12/10/1995
Hudgins, John, Jr., 3/18/1996
Hudson, M. Edward, 5/8/1997
Hughes, Howard A., 5/10/1990
Huisman, Marvin, 10/7/1999
Hummel, Joseph A., 1/16/1993
Humphrey, Robert Wilson, 9/17/2000
Husband, Paul Antonio Lyndell, Sr., 9/27/2000
Hyland, Neil, 3/3/1995
Hynes, James E., 10/27/1997
Insalaco, Leonard C., II, 3/15/1993
Jackson, Jimmy L., 9/18/1993
Jackson, Timothy Paul, 12/3/1999
Jackson, Robert Jeffery, 2/29/2000
Janora, David P., 1/2/1997
Jarvis, Joseph, Sr., 4/29/1994
Jenkins, Richard B., Sr., 8/22/1997
Jesse, Kenneth, 5/8/2000
Johnson, Rob, 7/6/1994
Johnson, Joe, 8/13/1994
Johnson, Robert L., 9/15/1994
Johnson, James H., 6/5/1997
Jones, William N., Jr., 11/22/1992
Jones, Hubert Sidney, 11/6/1998
Jones, Robert M., 1/12/2000
Joslyn, Daniel R., 6/1/1990
Kaczka, Donald, 3/24/1995
Kahn, Peter, 2/16/1997
Kail, Joseph P., 7/11/1991
Kalous, Richard L., 2/10/1998
Kaltreider, Ronald Eugene, 12/29/1999
Kaminski, Stanley F., 5/26/1997
Kane, Kevin C., 9/12/1991
Karl, George J., 7/17/1990
Karr, Arthur, 8/28/1993
Kautz, Martin, 8/31/1995
Keel, Kevin D., 9/18/1991
Keeth, Gifford, 7/26/1991
Keith, Craig W., 9/15/1991
Kelley., John L., Jr., 3/18/1990
Kelly, Robert, 7/29/1994
Kelso, Jon, 7/6/1994
Kennedy, John Rochford, 7/17/1998
Kennicutt, Gary L., 4/10/1993
Kerr, Kenneth W., 11/15/2000
Ketelsen,, Robert Boy, 1/27/2000
Key, Charles Franklin, 6/27/1998
Kibbey, Jennifer L., 3/13/1991
Kilgore, Walter, 1/5/1995
Kim, Henry Young Hi, 6/21/1991
King, Gary, 3/22/1994
King, John, 9/23/1994
King, Marcus, 1/31/1995
King, Patrick Joseph, 2/11/1998
King, Larry Joe, Sr., 8/16/1998
King, Daniel I., 11/26/2000
Kittle, Jimmy "Wayne", 2/12/1999
Klein, William T., 1/1/1990
Koebel, Donald, 3/8/1995

Kolenda, Marc, 2/14/1995
Konecny, Douglas K., 1/31/1993
Koponen, Leo, 5/15/2000
Korte, William Walter, 11/4/1999
Kozlowski, Joseph E., 8/27/1993
Kroboth, Joseph, Jr., 5/2/1998
Kroening, John F., 11/7/1997
Kucich, John A., 11/6/1991
Kuhn, Willard C., 1/28/1990
Labance, George S., 10/11/1990
Lafferty, Patrick, 10/13/1993
Lafon, James R., 8/24/1993
Lancaster, David Zan, 11/7/1999
Lane, Thomas L., 5/1/1990
Langley, Jeffrey, 3/30/1993
Langvardt, Mark W., 9/28/1992
LaPiedra, Scott J., 6/5/1998
Lapp, Robert, 5/13/1995
Lauber, Charles E. H., Jr., 12/17/2000
Laux, Paul Allen, 11/7/1998
Laverty, Scott T., 5/31/1991
Laws, Brett A., 2/6/1997
Leasher, Terry, 4/10/1996
LeClair, Richard J., 10/10/2000
LeClaire, Rebecca Ann, 12/1/1992
Lee, Robert E., 3/11/1990
Lee, James Hugh, 6/25/1991
Lee, Robert Odell, 9/30/1998
Lee, Brian Andrew, 11/16/1999
Lehman, Lawrence D., 7/7/1999
Lencioni, Ray, 9/19/1995
Lener, George, 6/5/1994
Lerma, Javier, 3/8/2000
Lewis, Johnny, Jr., 9/7/1991
Lewis, Tim L., 10/24/1992
Lewis, Alton L. "Al", 11/20/1999
Liddy, Richard, 11/9/1994
Lincoln, John F., Jr., 12/5/1997
Lindner, Bruce, 7/11/1996
Linkroum, Dale E., 12/6/1993
Liston, David John, 4/29/2000
Lockhart, Anthony E., 2/11/1998
Lohbeck, Michael, 4/17/1995
Lombardo, John F., 3/15/1993
Lopez, Ruben, 12/4/1996
Lorenzano, John J., 2/5/1992
Love, Charles, 1/10/1991
Lowery, Charles R., II, 10/20/1990
Loyd, Ralph J., 1/27/1999
Luby, Patrick J., 7/29/1992
Lucey, Jeremiah Michael, 12/3/1999
Lucier, Robert J., Sr., 8/19/1991
Luecht, Wayne Robert, 6/11/1999
Luker, William, 7/30/1995
Lumbra, David K., 8/9/1992
Lupo, Ronald, 6/24/1996
Luster-Stauss, Judith, 4/17/1995
Lynn, Randy, 7/29/1994
Lyons, Harold J., Sr., 10/30/1992
Lyons, James Francis, 12/3/1999
Mackey, Don, 7/6/1994
MacMurray, Paul, 8/27/1994
Maguidad, Leonardo, 3/2/1996

Maisano, David Anthony, 4/4/2000
Malone, Kevin, 5/26/1996
Mambretti, Louis, 3/9/1995
Manchester, Roc E., 11/6/1992
Mangieri, Eric, 8/7/1995
Manka, Norman, 3/13/1996
Manuel, Donald, 11/24/1996
Mapes, Michael D., 10/2/1997
Maplesden, Sydney Bruce, Jr., 8/25/1994
Margerum, Nelson, 3/15/1992
Maricle, Lynn E., 1/20/1991
Marks, William, 8/5/1995
Marshall, William H., Jr., 2/15/1991
Martin, William F., 5/31/1991
Martin, Leonard D., 10/1/1992
Martin, Carter, 6/22/1995
Martin, Allen H., Jr., 1/10/1997
Martin, Timothy Wayne, 6/9/1997
Martin, Donald Claude, 8/3/1998
Martinson, Robert William, Sr., 1/26/1997
Marvel, Carrol D., 1/15/1990
Masto, Stephen Joseph, 8/27/1999
Mathis, Michael, 4/11/1994
Mathis, Marvin, 1/31/1996
Matter, Edward J., Jr., 3/8/1998
Matthews, Lewis Jefferson, 5/30/1999
Maturen, Daniel L., 2/23/1992
Maxeiner, Thomas H., 12/23/1990
Maxwell, Elvis Benson, 10/7/1999
Mayfield, Steven C., 12/2/1998
Mayfield,, Kathryn A., 10/25/1997
Mayo, Lewis Evans, III, 2/14/2000
McAdams, William, 5/12/1990
McAllister, Phyllis, 2/24/1991
McAndrew, Ronnie M., 11/14/1990
McCarthy, Joseph P., 11/24/1991
McCarty, Sam, 8/18/1994
McClain, Curtis, 6/17/1990
McComb, Michael D., 3/23/1998
McCormack, Thomas M., 12/1/1997
McCormick, Travis, 5/9/1995
McCormick, Jared Conner, 11/2/2000
McCroden, John, 9/13/1995
McDonald, Randall P., 3/8/1991
McDonough, Eugene P., 9/5/1998
McElroy, Ernest Alan, 9/9/1998
McElveen, Terry, 10/27/1997
McEwen, Roger B., Sr., 7/4/1999
McGary, Dennis, 3/7/1996
McGinnis, William, 8/8/1996
McGowan, Harold "Mac" E., 1/2/1997
McGrath, Paul Francis, 6/17/1999
McGroarty, James H., 1/19/1999
McGuirk, Joseph T., 12/3/1999
McKay, Ray Parnell, Jr., 5/12/1995
McKenzie, Francis N., 1/18/1990
McKinney, Patrick Henry, Jr., 5/5/1998
McLaughlin, Peter, 10/8/1995
McLaughlin, Michael, 2/11/1996
McNally, David M., 12/22/1999
McNamee, Steven J., 11/23/1993
McNeer, Bradley Curtis, 12/18/1999
McNeil, Earl, Jr., 4/23/1995

McQuaide, Richard C., Sr., 2/18/1993
Medlicott, Robert, 7/29/1992
Meegan, Vincent D., Jr., 12/23/1993
Melendy, Victor, 1/28/1995
Mello, Donald V., 3/12/1991
Mellon, William C., 6/19/1997
Melloni, Richard H., Sr., 10/19/1993
Melton, Justin, 8/29/1998
Mendonsa, Tony F., 1/26/1993
Mercado, Jesus, 5/7/1998
Mercer, Larry L., 4/17/1997
Meredith, Kim, 7/15/1992
Merrell, L. C., 4/29/2000
Meshell, Ronald Wade, 8/8/1999
Metts, Brian E., 2/27/1993
Meyer, John J., 7/27/1990
Michael Smith, Michael, 6/22/1995
Michener, Brad A., 12/2/1999
Mikkelsen, Curtis D., 1/12/1991
Miller, Eugene, 3/13/1991
Miller, Daniel J., Jr., 12/6/1991
Miller, Guy C., 6/19/1992
Miller, Kenneth T., 10/16/2000
Million, Robert F., 2/15/1991
Mills, Leonard E., Sr., 7/28/1990
Mills, Dustin, 3/22/1994
Milner, Don L., 11/17/1992
Minehan, Stephen, 6/24/1994
Mitchell, Mark, 3/2/1994
Mondy, Delmar M., 7/27/1993
Moore, Mark A., 7/30/1990
Moore, Ricky G., 6/28/1997
Moore, Clifford Thomas, 6/16/1999
Moree, Don, 4/24/1996
Morgan, Newt, 2/7/1994
Morgan, Corey, 2/5/1996
Moriarty, Michael, 5/28/1991
Morris, Robert D., II, 6/29/1991
Morris, Calvin, 8/16/1992
Morrison, Dana, 3/27/1995
Morton, Reed, Sr., 1/19/1996
Morvant, Wylie J., 2/23/1991
Mosher, David, 1/3/1994
Mousley, Prince Albert, Jr., 1/6/1998
Mullins, Gerald, 1/8/1994
Mullins, Dennis, Jr., 1/12/1994
Mumford, Daniel W., 5/9/1998
Munter, Robert W., 5/29/1998
Murdock, Edward W., Sr., 1/10/1991
Murphy, Stephen Earl, 1/27/98
Murphy, Joseph R. "Dick", 1/29/1999
Murphy, John E., 4/8/1999
Murray, Gerald, 8/27/1994
Myers, Brian D., Sr., 1/1/1997
Myers, Warren D., Jr., 2/11/1998
Myers, Gerald "Jerry", 2/9/1999
Myers, Terry Lee, 2/15/1999
Nagel, Gary D., 10/5/1998
Nakovics, Raymond, 4/29/1998
Nall, David Thomas, 8/26/1999
Nam, Darrell R., 2/28/1992
Nason, Francis E., 11/5/1991
Navarro, John A., 10/29/1992

Negron, Carlos A., 3/20/1993
Nelboeck, Dale, 11/28/1994
Netsch, Lisa, 6/22/1995
Neuner, Michael E., Sr, 6/22/1997
Newcomb, Russell T., 12/14/1993
Newman, David Timothy, 9/7/2000
Nichols, Francis M., 1/23/1993
Nickell, Kenneth Allen, 4/6/1999
Nicosia, John A., 1/17/1991
Northam, William Jack, 6/16/1997
Novosad, Joe, 4/30/1995
Nutter, John, 8/3/1994
Oatman, Johnson "Jack", 11/6/1997
O'Boyle, Thomas, 9/26/1995
O'Brien, Shawn, 2/22/1995
O'Connell, Daniel J., 1/8/1999
O'Conner, John M., 2/1/1993
Ogbum, Warren R., 2/24/1993
Oliver, Josh, 7/28/1995
Oliver, Terry "Ted", 2/19/1999
Olsen, Carl Arnold, 1/5/1999
Olson, Dennis R., 3/6/1993
O'Rouke, Ronald L., 1/4/1993
Osadacz, Ronald J., 1/11/2000
Osmun, Jeffrey G., 12/5/1992
O'Toole, Robert J., 1/12/1998
Ottonello, Eugene, 11/13/1997
Ousley, George William, 5/18/1992
Ousley, Earnest Otis, 9/4/2000
Overman, William M., Jr., 4/24/1993
Ovsiovitch, Elias, 10/1/1994
Pache, John, 9/8/1995
Pacheco, Jesse, Jr., 12/6/1993
Pacheco, Gregory Edwin, 10/3/1999
Pack, David Merle, 10/29/1999
Packard, David L., 3/16/1999
Paris, Daniel G., 6/19/1991
Park, Johnnie Ray, 6/27/1998
Parker, Robert L., 3/30/1991
Parker, Tommy A., 11/4/1992
Parks, David Vernon, 7/12/1999
Parris, Cogdil Jacob, Jr., 5/21/1993
Parsons, William L., 7/4/1996
Passaro, Gary M., 1/18/1990
Patterson, Preston Edgar, 9/27/1998
Payne, Carl Ray, 5/7/2000
Payton, Donald J., Sr., 9/8/1997
Pearson, Arthur Lo, 10/25/1992
Pemberton, Robert, 4/26/1996
Penning, Richard A., 11/5/1990
Perry, Aaron J., 6/27/1990
Perry, Henry E., 9/14/1997
Pescatore, Nathan Andrew, 7/2/2000
Peters, George, 9/6/1995
Peters, Robert F., 8/29/1998
Peters, Robert D., 4/15/1999
Petersen, Lance J., 11/1/1990
Peterson, Clyde, 6/8/1999
Petit, Arthur, 7/13/1996
Phillips, Anthony Sean, Sr., 5/30/1999
Phillips, Ronald Gregory, 6/18/1999
Pierce, Everett C., 8/15/1993
Pierce, Bradley Scott, 8/6/2000

Pinkowski, Phillip M., Jr., 4/12/1999
Pinnell, Wilbur, 1/7/1995
Pitcher, Edward, 7/15/1995
Pizinger, Michael A., 5/2/1998
Ploeger, Francis, 2/24/1996
Plummer, John N., 1/27/1991
Polan, Mark E., 8/22/1990
Pollard, Guy R., 1/6/1996
Pollard, Robert Dale, 12/31/1999
Porter, Gary S., 10/24/1992
Post, Guy E., 11/30/1993
Pottberg, Bryan Christopher, 7/15/1999
Powell, Billy Ray, 3/23/1992
Powers, Louis T., 8/5/1993
Prendergast, Thomas E., 7/23/1998
Price, Floyd M., 3/18/1990
Prime, Norman, 3/29/1995
Pritchett, John Paul "J.P.", Sr., 8/20/2000
Pucker, Dean O., 7/31/1992
Pulley, Wayne L., 8/21/1991
Purcell, Keith P., 12/17/2000
Purdy, Lee A., 1/8/2000
Quadrel, Frank A., 1/1/1991
Queen, Michael "Mike" Russell, 3/28/2000
Queen, Elwood, 11/29/2000
Quinn, Nathaniel, 2/23/1996
Quinn, Gregory I., 11/22/1997
Quinones, Jaime, Jr., 8/26/2000
Raibley, Donald, 7/21/1996
Ramey, Jerry Wayne, 11/3/1999
Ramos, Edward, 11/27/1996
Raskin, Daniel J., 7/9/1990
Rathbun, Albert Roger "Bo", 10/8/2000
Ray, James R., 5/15/1990
Ray, Norman, 8/8/1996
Ray, David M., 5/29/1997
Reavis, James, 10/26/2000
Redmond, John, 1/28/1994
Reed, Jerry A., 5/28/1990
Reick, Jeffrey William, 4/1/1998
Reid, William, 9/20/1996
Reiner, Eric F., 3/23/1998
Reits, Robert G., Sr., 1/16/1992
Renfroe, James Robert, 8/14/2000
Renner, Jeffrey, 9/18/1996
Reveal, Kevin, 9/29/1996
Reynaga, Randy W., 7/3/1993
Reynolds, Thomas C., 1/20/1997
Rhinehart, Bruce A., 10/27/1990
Rhoads, Stanley, 4/15/1994
Rice, Mark S., 12/1/1991
Rice, Richard K., Sr., 3/28/1998
Rice, Thomas Benjamin, 11/9/1998
Richter, Karl, 6/14/1990
Ridings, Phillip, 7/15/2000
Riggins, John, Jr., 10/25/1995
Riley, James M., Jr., 10/20/1991
Rivera, Heriberto T., 5/10/1990
Rivera-Rivas, Luis A., 2/5/1998
Rizac, Christopher, 6/10/1995
Robbins, Rick, 4/24/1996
Robertson, William J., 4/25/1998
Robibero, Albert F., 6/24/1991

Robinson, Reginald G., Sr., 5/8/1997
Robinson, Costello Nathaniel "Colonel", 7/7/1999
Roche, Lawrence, 8/26/1996
Rodd, Dennis, 3/13/1993
Rodgers, Gregory Eugene, 12/13/1999
Rodriguez, Evangelino Soto, 2/21/2000
Roemer, Harold E., Jr., 1/5/1998
Rohrbaugh, Douglas L., 6/26/1998
Ronaldson, Alfred E., 3/5/1991
Rosen, Wayne, 6/22/1999
Roth, Roger, 7/6/1994
Rovero, Malcolm A., 7/15/1997
Rowe, Will Ellis, Jr., 5/9/1997
Rudd, Charles A., 6/5/1997
Ruezga, Arthur, 8/20/1993
Russ, Edward A., 12/11/2000
Russell, Wayne Ronald, 7/6/1991
Ruth, Victor, III, 6/10/1994
Rutledge, Christopher C., 10/22/1993
Ryan, Thomas P., 12/30/1997
Sadowski, Peter S., 1/29/1993
Salisbury, Harold, 1/10/1994
Samanas, Robert M., 10/31/2000
Sammons, Jeffrey E., 8/19/1997
Sanders, Richard, 1/15/1997
Sanders, Gary Charles, 2/15/99
Sapp, James D., 9/2/1991
Sarles, Marshall F., 7/2/1991
Satterfield, Paul P., 9/28/1998
Savage, Christopher P., 2/27/1993
Savage, Karen Jane, 10/16/1999
Scannell, Bernard D. (Pete), 9/21/2000
Scarborough, Timothy P., 4/15/1991
Schaefer, Eric, 9/16/1995
Schiebel, Raymond, 3/5/1995
Schmidt, Gordon O., 11/11/1991
Schmidt, Karl, 10/15/1996
Schmitt, Howard R., 1/29/1993
Schott, James W., 1/5/1992
Schubert, Gene, 9/17/1995
Schumacher, Arthur E., 12/31/1992
Schunk, Norman G., Jr., 1/9/1993
Schuyler, John, 7/10/1995
Schwinger, Walter. Jr., 11/12/1996
Scott, Glenn, 2/8/1995
Scott, Henry, 9/18/1996
Scott, Stanley, 12/21/1996
Seaburg, Kevin C., 2/5/1997
Seguin, Michael L., 7/4/1997
Segura, Hector M., 5/23/1990
Seib, Dale M., 2/26/1990
Seidenburg, Christopher, 3/29/1994
Selby, Scott, 8/29/1998
Sexton, Lydia A., 10/24/1990
Shadrick, Lester Lee, 8/13/2000
Sharp, David Clements, II, 3/17/2000
Shaughnessy, Michael, 7/23/1994
Shaw, Robert A., 10/16/1991
Sheats, Lewis R., 4/1/1993
Shenefield, Ray J., 12/30/1990
Sheppard, Ann F., 2/20/1994
Sherburn, Phillip, 3/15/1995
Sheridan, Charles F., 10/1/1992

Sherman, Kris, 7/27/1996
Shirk, Evan N., 5/27/2000
Shockley, Jesse, Jr., 12/10/1994
Shoemaker, Gregory, 1/5/1995
Shortt, Mike, 3/15/2000
Shue, James, 12/14/1995
Sieglinger, John R., 10/16/1991
Sienknecht, John H., 12/8/1991
Simmons, Nonnan L., 10/7/1991
Simpson, Edwin L., 7/14/1990
Sims, Randy, 9/14/1998
Sims, Michael J., Sr., 11/2/1999
Sippel, Albert, 7/16/1997
Siverton, Walter E., Jr., 1/8/1993
Sligar, Arch Russell, Jr., 6/13/1999
Smartt, "Randy", 12/9/1997
Smith, Albert R., 12/3/1991
Smith, Frank Albert, 5/26/1992
Smith, Sam, 7/12/1994
Smith, Wayne, 8/7/1994
Smith, Herbert, 8/18/1994
Smith, Dwight, 9/6/1994
Smith, Daren, 10/6/1994
Smith, Lathan Grant, Jr., 1/26/1995
Smith, Michelle, 6/9/1996
Smith, Kevin Rex, 4/6/1999
Smith, Kimberly Ann, 2/14/2000
Smith, Lyndell J., 5/31/2000
Smith, Warren (J. C.), 8/13/2000
Smith, Phillip Dewey, 11/16/2000
Smitherman, William "Sam", Sr., 12/20/1997
Snell-Dean, Phillip Wayne, 2/15/99
Somay, John H., 11/12/1993
Soper, Edward E., 1/26/1991
Sorenson, Adam, 7/15/1995
Soupene, Gary, 7/10/1995
Sowle, Ingrid H., 2/19/1990
Spangler, John E., 10/29/1991
Spencer, John H., 4/14/1991
Spencer, Thomas Edward, 12/3/1999
Spink, Richard Owen, 2/13/2000
Springfield, Curtis E., 6/26/1990
Stanbery, Ralph William, 4/22/1998
Stanmire, Robert C., Sr., 2/16/1999
Stark, Roger, 6/19/1992
Stark, Richard, 8/2/2000
Starr, Julius C., 8/15/1990
Stavely, James H., III, 6/30/1992
Steele, Lee Allen, 6/23/1996
Stephen E. Bovey, Ralph F., 9/30/1990
Stephenson, Roy, 10/13/1994
Sterenchuk, Kevin Francis, 3/27/2000
Stevens, Leo A., 6/3/1997
Stevens, Douglas George, 2/6/2000
Stiles, Martin Michael, 7/18/1999
Stine, Timothy M., 11/13/1990
Stivers, Dania, 5/12/1995
Strain, Kenneth Alan, 5/2/1999
Strall, Sam, 9/14/1996
Straub, James D., 3/26/1990
Streeter, Allen L., 1/11/2000
Stroud, Ronald L., 1/29/1990
Strube, Howard E., 9/7/1997

Struble, L. Wayne, 4/13/1991
Sulzinski, Stephen, 10/29/995
Sutch, Kevin, 6/12/1995
Sutton, Brian D., Sr., 11/8/1994
Sutton, David Paul, 3/4/2000
Swan, John William, II, 8/6/1996
Swenson, Leroy, 11/2/1997
Swindle, James T., 7/5/1991
Swinehart, Roy J., 4/21/1992
Tagliareni, Joseph F., Jr., 5/31/1999
Talley, Raymond E., 1/18/1992
Taylor, Jerome, 3/12/1999
Tebo, James H., 7/13/1997
Teehan, Whitney C., Jr., 6/25/2000
Templin, Frederick W., 10/14/1991
Terlicker, Randall, 1/5/1995
Theisen, David P., 2/5/1998
Thomas, Keith C., 2/17/1998
Thomason, Curtis C., 4/22/1990
Thompson, Arthur, 7/19/1995
Thompson, Jeffrey Scott, 10/4/1999
Thorn, Glen, 1/21/1994
Thrash, James, 7/6/1994
Thrower, Lawrence D., 10/24/1998
Tillman, Aubrey R., 4/2/1999
Tippins, Steve Austin, 12/9/1998
Tobias, Samuel James, 5/15/2000
Tolan, James R., 1/16/1999
Toledo, Frankie, 4/22/1993
Toomey, Tracy Dolan, 1/10/1999
Trice, Steve, 11/9/1996
Trick, Grant F., 8/13/2000
Trotochaud, Donald, 9/18/1998
Trygar, Raymond, 8/20/1995
Tuck, Arthur K., 3/13/1992
Tuck, Nathan R., 12/22/1999
Tullis, Arthur A., 5/4/1999
Tvedten,, John H., 12/20/1999
Tyler, Richard, 7/6/1994
Ulrich, Robert Charles, 10/28/1999
Vagnier, Joseph M., 7/1/1997
Valdez, Roberto, 1/29/1992
Valentino, Louis, 2/5/1996
Van Calbergh, Michael E., 6/30/1990
Van Wert, Richard L., 12/10/1999
VanAuken, Gail Lynne, 11/2/2000
Vanhoy, Howard William, 9/5/2000
Vaughan, Walter F., 10/29/1999
Veri, Frank, Jr., 12/20/1991
Viloria, Marshal E., 3/1/1991
Vinson, Raymond, 2/11/1996
Vodak, Charles James "Chuck", 3/16/1999
Voris, Albert Leonel, Jr., 9/1/2000
Vrabel, Scott Alan, 11/14/1997
Vreeland, Rick A., 2/5/1993
Wade, Walter, 2/6/1994
Walker, Wayne E., 6/8/1992
Walker, Keith A., 3/7/1993
Walker, John M., 8/20/1998
Wall, Thomas Oscar, 10/5/1998
Walling, James D., 12/9/1991
Wallingford, Robert, 8/27/1996
Walls, Nathan E., 10/11/1991

Walls, William, 6/5/1995
Walsh, Larry R., 4/9/1998
Wannagot, Daniel E., 3/15/1991
Ward+C406ell, Maurice, Jr., 1/23/1994
Warick, James, 12/23/1996
Warner, Chester F., 10/1/1992
Warren, Alton, 5/30/1994
Warren, Timothy J., III, 2/15/1997
Warren, Stewart, 5/8/1997
Wary, Barry L., 10/13/1998
Washburn, Richard, 9/30/1995
Waskiewicz, Robert, 4/3/1994
Waters, Roland, 6/14/1992
Watson, John McClay, 6/16/1997
Watts, David J., 4/28/1999
Wauson, Martin Richard, 1/11/1999
Weaver, James, 4/2/1995
Weaver, Mitch, 7/25/1995
Weber, Charles Allen, Sr., 2/28/1997
Weeks, Thomas, 8/25/1990
Weeks, Bryan Lo, 7/12/1992
Weingart, John, 7/14/1995
Weischman, George Delbert, 12/28/1992
Wheeler, William M , 1/2/1993
White, James E., 1/10/1990
Whiteside, George R., 12/13/1991
Wiborg, Michael Curtis, 4/13/1998
Wiebe, Robert P., 9/11/1990
Wilcom, Michael J., Jr., 5/24/1993
Wilkes, Mark A., 9/2/1991
Williams, Gregory E., 11/4/1991
Williams, Thomas A., 2/24/1992
Williams, Marilyn, 2/1/1994
Williams, Henry, 3/18/1995
Williams, James B., 1/5/1996

Williams, Charles "Chuck" H., II, 2/17/1997
Williams, David Shawn, 10/22/1997
Williams, Kennon Loy, 12/31/1998
Williams, Lewis Edward, 5/14/1999
Williford, Randy, 7/6/1995
Wilmot, Steven Max, 7/18/2000
Wilson, Richard E., 6/29/1992
Wilson, Ronnie, 4/15/1995
Wilson, William "Junior" T., 5/25/1997
Wilson, Donald R., 3/6/2000
Wilt, Joseph M., 1/1/1990
Winckler, George S., 10/18/1991
Winters, James C., 12/20/1990
Winters, William "Pops" H., 10/10/1997
Womer, David, 10/24/1997
Wood, Frank William, 8/14/1999
Woodward, John, 6/29/1995
Wright, Blake V., 8/26/1990
Wright, Jared Lee, 1/31/1995
Wunch, Mark A., 5/13/1990
Wylie, Thomas, 12/27/1994
Yahraus, Michael Kenneth, 9/11/2000
Yale, Stephen D., 12/20/1991
Yankey, William Dwight, 11/13/1998
Yanklin, Daniel H., 9/4/2000
Y'Barbo, Jessie Lamar, 3/13/2000
Yost, Wayne C., 11/20/1999
Young, Julie Ann, 7/30/1992
Young, James, 3/29/1994
Young, Frank, 3/18/1996
Young, Charles C., 10/18/1999
Young, Ernest John, 1/16/2000
Zaremba, Shawn, 8/13/1994
Zeller, Leonard N., 12/15/1997
Zimmerman, Dale, 1/13/1996